T0196101

MACAT

An Analysis of

Mathis Wackernagel and William Rees's

Our Ecological Footprint
Reducing Human Impact on the Earth

Luca Marazzi

Published by Macat International Ltd
24:13 Coda Centre, 189 Munster Road, London SW6 6AW.

Distributed exclusively by Routledge
2 Park Square, Milton Park, Abingdon, Oxon OX14 4RN
711 Third Avenue, New York, NY 10017, USA

Routledge is an imprint of the Taylor & Francis Group, an informa business

www.macat.com
info@macat.com

Cataloguing in Publication Data
A catalogue record for this book is available from the British Library.
Library of Congress Cataloguing-in-Publication Data is available upon request.
Cover illustration: Alex Tomlinson

ISBN 978-1-912302-40-6 (hardback)
ISBN 978-1-912128-02-0 (paperback)
ISBN 978-1-912281-28-2 (e-book)

Notice
The information in this book is designed to orientate readers of the work under analysis,
to elucidate and contextualise its key ideas and themes, and to aid in the development
of critical thinking skills. It is not meant to be used, nor should it be used, as a
substitute for original thinking or in place of original writing or research. References and
notes are provided for informational purposes and their presence does not constitute
endorsement of the information or opinions therein. This book is presented solely for
educational purposes. It is sold on the understanding that the publisher is not engaged
to provide any scholarly advice. The publisher has made every effort to ensure that
this book is accurate and up-to-date, but makes no warranties or representations with
regard to the completeness or reliability of the information it contains. The information
and the opinions provided herein are not guaranteed or warranted to produce particular
results and may not be suitable for students of every ability. The publisher shall not be
liable for any loss, damage or disruption arising from any errors or omissions, or from
the use of this book, including, but not limited to, special, incidental, consequential or
other damages caused, or alleged to have been caused, directly or indirectly, by the
information contained within.

CONTENTS

THE MACAT LIBRARY

The Macat Library is a series of unique academic explorations of seminal works in the humanities and social sciences – books and papers that have had a significant and widely recognised impact on their disciplines. It has been created to serve as much more than just a summary of what lies between the covers of a great book. It illuminates and explores the influences on, ideas of, and impact of that book. Our goal is to offer a learning resource that encourages critical thinking and fosters a better, deeper understanding of important ideas.

Each publication is divided into three Sections: Influences, Ideas, and Impact. Each Section has four Modules. These explore every important facet of the work, and the responses to it.

This Section-Module structure makes a Macat Library book easy to use, but it has another important feature. Because each Macat book is written to the same format, it is possible (and encouraged!) to cross-reference multiple Macat books along the same lines of inquiry or research. This allows the reader to open up interesting interdisciplinary pathways.

To further aid your reading, lists of glossary terms and people mentioned are included at the end of this book (these are indicated by an asterisk [*] throughout) – as well as a list of works cited.

Macat has worked with the University of Cambridge to identify the elements of critical thinking and understand the ways in which six different skills combine to enable effective thinking.
Three allow us to fully understand a problem; three more give us the tools to solve it. Together, these six skills make up the **PACIER** model of critical thinking. They are:

ANALYSIS – understanding how an argument is built
EVALUATION – exploring the strengths and weaknesses of an argument
INTERPRETATION – understanding issues of meaning

CREATIVE THINKING – coming up with new ideas and fresh connections
PROBLEM-SOLVING – producing strong solutions
REASONING – creating strong arguments

To find out more, visit **WWW.MACAT.COM.**

CRITICAL THINKING AND *OUR ECOLOGICAL FOOTPRINT*

Primary critical thinking skill: PROBLEM-SOLVING
Secondary critical thinking skill: CREATIVE THINKING

Our Ecological Footprint is a masterclass in the critical thinking skill of problem-solving, with a good dose of creative thinking added to the process.

Geographers, ecologists and environmentalists have long been concerned about the speed at which the human race is consuming Earth's natural resources. What they did not have, however, was a reliable way of measuring ecological impact in the first place. Wackernagel and Rees identified and defined this as a clear problem for any impact-reduction strategy. They then set about resolving the difficulty by creating a suitable methodology for measuring the impact of human actions on the planet. That methodology identified relevant criteria and combined them into one calculation, expressing environmental impact in terms of the area of land needed to support our needs.

Not only was this highly creative (nobody had expressed impact in terms of land usage before), it also provided an instantly comprehensible way of assessing environmental impact, allowing scientists and policymakers to test much more effectively any strategy that might reduce it. Giving individuals and governments alike a direct and simple measure of their impact, the 'footprint' method is now used all over the world to develop strategies for saving the planet's resources.

ABOUT THE AUTHORS OF THE ORIGINAL WORK

Born in 1962, **Mathis Wackernagel** is a Swiss-born academic and advocate for sustainability. He studied for his PhD in community and regional planning at the University of British Columbia in Canada under the supervision of the population ecologist—and his future co-author—William Rees.

Born in 1943, **Rees** is a Canadian professor of community and regional planning, with decades of experience studying the limits of the Earth's resources, and the consequences of modern lifestyles that often demand resources and create waste beyond the Earth's capacity to cope.

Following the success of their co-authored 1996 book, *Our Ecological Footprint*, both men have gone on to found nongovernmental organizations (NGOs) that design and implement strategies to help foster a cleaner environment and fairer societies around the world.

ABOUT THE AUTHOR OF THE ANALYSIS

Dr Luca Marazzi holds a PhD in freshwater ecology from the Environmental Change Research Centre at University College, London. He is currently undertaking postdoctoral research at the Southeast Environment Research Center, Florida International University.

ABOUT MACAT

GREAT WORKS FOR CRITICAL THINKING

Macat is focused on making the ideas of the world's great thinkers accessible and comprehensible to everybody, everywhere, in ways that promote the development of enhanced critical thinking skills.

It works with leading academics from the world's top universities to produce new analyses that focus on the ideas and the impact of the most influential works ever written across a wide variety of academic disciplines. Each of the works that sit at the heart of its growing library is an enduring example of great thinking. But by setting them in context – and looking at the influences that shaped their authors, as well as the responses they provoked – Macat encourages readers to look at these classics and game-changers with fresh eyes. Readers learn to think, engage and challenge their ideas, rather than simply accepting them.

'Macat offers an amazing first-of-its-kind tool for interdisciplinary learning and research. Its focus on works that transformed their disciplines and its rigorous approach, drawing on the world's leading experts and educational institutions, opens up a world-class education to anyone.'

Andreas Schleicher
Director for Education and Skills, Organisation for Economic Co-operation and Development

'Macat is taking on some of the major challenges in university education … They have drawn together a strong team of active academics who are producing teaching materials that are novel in the breadth of their approach.'

Prof Lord Broers,
former Vice-Chancellor of the University of Cambridge

'The Macat vision is exceptionally exciting. It focuses upon new modes of learning which analyse and explain seminal texts which have profoundly influenced world thinking and so social and economic development. It promotes the kind of critical thinking which is essential for any society and economy.
This is the learning of the future.'

Rt Hon Charles Clarke, former UK Secretary of State for Education

'The Macat analyses provide immediate access to the critical conversation surrounding the books that have shaped their respective discipline, which will make them an invaluable resource to all of those, students and teachers, working in the field.'

Professor William Tronzo, University of California at San Diego

WAYS IN TO THE TEXT

KEY POINTS

- Born in 1943, William Rees is a Canadian population ecologist* and professor emeritus at the University of British Columbia. Population ecology is a subfield of ecology* concerned with the study of the distribution, dynamics, and structure of animal and plant populations. Born in 1962, Mathis Wackernagel is Rees's former PhD student in community and regional planning,* an academic and professional field concerned with helping communities develop the local economy and infrastructure. Together, Rees and Wackernagel developed and coauthored 1996's *Our Ecological Footprint*.

- An "ecological footprint"* is the amount of biologically productive land required to sustain a population's consumption of natural resources and dispose of their waste products. Using this concept, the authors challenge traditional economic models and propose strategies to promote a cleaner and fairer global environment.

- Initiatives to reduce the size of our ecological footprints have flourished all over the world. These include projects undertaken by the authors' nongovernmental organizations, the Global Footprint Network* and the One Earth initiative.*

Who Are Mathis Wackernagel and William Rees?

Mathis Wackernagel, coauthor of *Our Ecological Footprint: Reducing Human Impact on the Earth* (1996), is a sustainability planner*— someone who devises strategies to help develop infrastructure and economic models that focus on making sure resources will continue to be available to future generations. He is also the cofounder and president of the Global Footprint Network. Born in Basel, Switzerland in 1962, Wackernagel earned a degree in mechanical engineering from the Swiss Federal Institute of Technology in 1987 and a PhD in community and regional planning at the University of British Columbia (UBC) in Canada in 1994.

William Rees, the book's second author, was born in Brandon, Canada in 1943. A population ecologist who earned his PhD from the University of Toronto in 1969, he is now professor emeritus at UBC's School of Community and Regional Planning. Rees supervised Wackernagel's doctoral dissertation, and as a result of this close work together on shared academic and social interests, they continued to collaborate as colleagues.

Their 1996 coauthored work *Our Ecological Footprint* was a "collateral" outcome—a by-product—of Wackernagel's thesis, which first developed the concept and calculation methodology known as the ecological footprint. In the book, the authors elaborate and refine the idea of an ecological footprint and present it in a way that allows responses from a wide audience.

The ecological footprint is designed to provide a way both to measure and to reduce ecological impacts on the Earth's finite stock of resources. Wackernagel and Rees focus on ways to quantify and represent human consumption of resources and the amount of land required to sustain this consumption. The book's straightforward approach allows ordinary people to use its concepts in their daily lives, and has made Wackernagel and Rees's ideas part of the ongoing academic and public debates about resource use and sustainability* (issues surrounding our need for economic development in a world of finite resources). In this context, sustainability refers to strategies to use natural resources in a way that allows them to continue to renew themselves. Preventing over-fishing so that fish stocks can replenish is a good example of this. After *Our Ecological Footprint* was published, the authors founded the Global Footprint Network and the One Earth initiative. These nongovernmental organizations (NGOs) design and implement strategies to reduce the ecological footprints of different communities around the world.[1]

What Does *Our Ecological Footprint* Say?

Our Ecological Footprint argues that modern societies have overconsumed and damaged the Earth's natural resources. It offers a method to quantify the impact of a range of human activities on the environment. Traditionally, ecologists have employed the concept of carrying capacity*—the maximum population size of a species that can be supported by an area while maintaining its capacity to do so in the future. However, Wackernagel and Rees took this formula (effectively, the number of people divided by the land area) and turned it the other way round to establish the concept of the ecological footprint: the amount of land needed to support a given population (the land area divided by the number of people).

This deceptively simple rearrangement produced a clear and useful indicator of the amount of productive land required to sustain an individual or a group at a given level of consumption. The straightforward nature of the concept and its descriptive power have helped to raise awareness about the ongoing exploitation of natural resources and made it a very popular planning and educational tool.

The authors deliberately confront the dominant neoliberal* economic model, which does not account for the Earth's limited resources. Neoliberalism is a political and economic doctrine in favor of privatization, free trade, reduced government interference in

11

business, and low public expenditure on social services. Economic growth is central to this model; *Our Ecological Footprint* shows that the idea of constant economic growth may not be possible in a world of finite natural resources.

Wackernagel and Rees helped to construct an alternative economic framework designed both to assess more accurately and to incorporate levels of consumption of both materials and energy, as well as waste production. The ecological footprint is a measure of how much land is needed to maintain a community's well-being, and the authors argue that land requirements provide an important measurement to help identify and address inequalities across societies, while also sustaining other species. The central idea is that individuals, communities, and even nations can employ the ecological footprint to first quantify, and then appropriately decrease, the consumption of material goods and energy to keep consumption in line with what the environment is able to provide without disastrous consequences.

The authors lay out their vision for a sustainable future that identifies excesses in consumption, and examines the limits to economic growth and the resources on which that growth is dependent. This is important, because it involves a fundamental rethinking of the very economic and behavioral foundations of most modern societies. Wackernagel and Rees argue that while human beings *can* live inside nature's limits, first we have to acknowledge and define these limitations. Establishing a means to estimate and communicate those limits is what *Our Ecological Footprint* really does.

The book is still extremely relevant today. According to the educational web browser Google Scholar, the text has been cited over 6,600 times.[2] By setting out their ideas, Wackernagel and Rees began to develop a practical sustainability program that can help communities at various levels of development and organization contribute to a new world order. This new world order offers trade that is ecologically sustainable, while consumption by the rich is limited to leave ecological space for the improvement of living conditions in poorer communities.

The authors' ambition, however, goes even further than proposing a revision of the global distribution of material wealth to include less tangible developmental goals. In fact, Wackernagel and Rees believe that their "brave new sustainable world will also be the kind of society that satisfies people's non-material needs."[3] In their view, quantifying (measuring) resource demands using the ecological footprint tool acknowledges human dependency on ecosystems* (that is, the complex set of interrelationships between and among a biological community and its physical environment), identifies areas where resources are over-exploited, and encourages both social and economic change. Although the ecological footprint tool recognizes fundamental limits to economic growth, it does not rule out social and economic development *within* those limits.

Why Does *Our Ecological Footprint* Matter?

Our Ecological Footprint matters because ecological overshoot*—the idea that natural resources are not replenished as fast as they are consumed—is an extremely dangerous situation for human beings and all of nature alike.[4] Exceeding the Earth's ecological limits puts other species in danger, and can result in vital resources disappearing or species becoming extinct. The book is a great resource for describing sustainability issues to a general audience, and has stimulated a productive debate about how to best measure the wide range of human impacts on the Earth. *Our Ecological Footprint* also highlights the unequal distribution of natural resources and demonstrates how the affluence of some communities is intimately tied to resource exploitation in poorer ones. The authors argue that international trade has established what is called a zero-sum* game: as some countries get materially richer, that means there are fewer resources available for others. To put these inequalities right and to reduce wealthier nations' over-exploitation of resources in poor countries, Wackernagel and Rees support a more mindful and localized production of goods and services to improve quality of life everywhere.

The authors anticipated that critics might challenge the somewhat ideological tone of their work, and so in the book they stress the role of technological efficiencies to help resolve over-exploitation and the unfair distribution of natural resources in the current economic and trading setups. They argue that global sustainability can only be achieved by changing the neoliberal economic system that currently dominates.

Wackernagel and Rees's principal argument can be summarized as follows: the free-trade/market-based economic system (an economic system in which government economic interference is minimal in order to allow the market to operate without hindrance) has inbuilt flaws that encourage both the accumulation of material wealth and excessive consumption by relatively few people. Much of the world's population, on the other hand, lacks the resources to meet basic human needs. This inequality also creates a disproportionate environmental impact, and those in the West have a *moral* duty to reduce this impact by consuming less, sharing more, and constructing a more just economic system. The authors argue that the solution needs engagement at all levels of society to acknowledge and define the planet's natural limits, and a reassessment of how we consume its resources and how we manage our waste.

Wackernagel and Rees's program has been successful from an academic point of view and has contributed to the development of the field of ecological economics,* a discipline that bridges not only ecology and economics, but also psychology, anthropology, archaeology, and history. Measuring its practical success, however, is more difficult. *Our Ecological Footprint* has certainly raised the consciousness of people everywhere about this issue. The nongovernmental organizations that Wackernagel and Rees have set up offer the tantalizing possibility of stimulating wider political and social change. As a more straightforward accounting tool, the concept of the ecological footprint has been well received and widely adopted. However, so far the associated social and political changes recommended by Wackernagel and Rees to reduce our ecological footprints have not been forthcoming. It seems there is a lack of both political will and public demand to implement them.

NOTES

1   "Home," Global Footprint Network, accessed January 24, 2016, http://footprintnetwork.org/; One Earth, "Projects," accessed January 24, 2016, http://www.oneearthweb.org/projects.html.

2   "Our Ecological Footprint," Google Scholar, accessed January 24, 2016, https://scholar.google.co.uk/scholar?hl=en&q=Our+Ecological+Footprint+&btnG=&as_sdt=1%2C5&as_sdtp=.

3   Mathis Wackernagel and William Rees, *Our Ecological Footprint: Reducing Human Impact on the Earth* (Gabriola Island, British Columbia: New Society Publishers, 1996), 79–120.

4   One Earth Web, "Rethinking your life," accessed September 10, 2012, http://oneearthweb.org/category/news/.

SECTION 1
INFLUENCES

MODULE 1
THE AUTHORS AND THE
HISTORICAL CONTEXT

KEY POINTS

- *Our Ecological Footprint* remains a relevant and practical tool to assess communities' ecological impacts at different scales, and continues to contribute to environmental policy debates.

- Mathis Wackernagel and William Rees's academic work was profoundly influenced by their formative experiences in Switzerland and Canada, respectively, and by significant milestones in the environmental movement from the 1960s onwards.

- Following the success of the concept of the ecological footprint*—the amount of productive land required to sustain communities' levels of resource consumption and to take care of its waste—the authors achieved important academic and public policy goals, including the establishment of influential nongovernmental organizations (NGOs).

Why Read This Text?

Mathis Wackernagel and William Rees's *Our Ecological Footprint: Reducing Human Impact on the Earth* (1996) is essential reading for the ever-growing number of people—in academia, policy-making roles, and the general public—interested in environmental sustainability* (our capacity to square the economic needs of human populations with the world's finite resources). It provides a straightforward approach to quantify the impact our activities have on nature, and makes recommendations about how we can reduce this impact. *Our*

> 66 On a finite planet, at human carrying capacity, a society driven mainly by selfish individualism has all the potential for sustainability of a collection of angry scorpions in a bottle. 99
>
> Wackernagel and Rees, *Our Ecological Footprint: Reducing Human Impact on the Earth*

Ecological Footprint provides a useful conceptual basis for debates on how to create a more sustainable economic model and more equal societies; it also sets forth an accounting tool we can use to work toward this goal. The authors argue that the neoliberal* economic system—an economic system defined by unhindered free trade—is inherently flawed because it encourages the accumulation of material wealth and overconsumption by a relatively small number of people, while the majority of the world's population has resources that are relatively scarce and too often unable to meet basic human needs.

The authors start with a basic but crucial point: an acknowledgment that the planet's natural resources are limited and that human societies are dependent on the Earth's capacity to regenerate resources. Their core concept, the ecological footprint, assesses the amount of productive land required to sustain communities' levels of resource consumption (such as food, electricity, and fuel) and to take care of its waste (such as landfill space and forests where the trees absorb some carbon emissions). This gave existing movements aimed at addressing environmental issues a numbers-based way of illustrating the strain human behavior puts on the planet.

Despite some early criticisms, this book has become a successful educational and civic planning tool. The ecological footprint model has continued to evolve as scholars who share Wackernagel and Rees's concerns have built on and refined it, and the work continues to provide a provocative basis for academic and public discussion.

Authors' Lives

Mathis Wackernagel was born in Basel, Switzerland in 1962. In a 2007 interview, he recalled a conversation he had with his grandfather in 1973 as the moment that opened his eyes to the limitations of the Earth's most vital resources. His grandfather was adamant that humans would be able to discover and exploit more oil indefinitely, but Wackernagel was struck that he, a "naive child," realized better than a "sophisticated adult" that nonrenewable energy resources—such as fossil fuels*— would eventually run out.[1] This seemingly ordinary encounter between a grandfather and grandson set the stage for his career-long interest in ecology* and sustainability. Wackernagel is married to Susan Burns,* the chief executive officer of the international sustainability think tank called the Global Footprint Network which was cofounded by Burns and Wackernagel.* Burns has almost two decades of experience advising global corporations and organizations on environmental sustainability.

William Rees was born in 1943 in Brandon in the Canadian province of Manitoba, and spent much of his childhood on his grandfather's farm on the St. Lawrence River, between Iroquois and Morrisburg, Ontario in Canada.[2] The natural environment that surrounded him when he was a child profoundly influenced his studies and professional work. In a 2000 interview with the journal *Aurora,* he revealed that it was during a family meal when he was nine years old that he first understood his profound connection between what he consumed and the land on which it was produced:"I realized that there wasn't a single thing on the plate that I hadn't had a hand in growing. I suppose it was like an epiphany kind of experience."[3] After receiving a PhD in population ecology at the University of Toronto in 1969, Rees became a professor at the University of British Columbia's School of Community and Regional Planning,* where he is currently professor emeritus. Like his coauthor, Rees founded an NGO—One Earth*— following the publication and success of *Our Ecological Footprint.*

From the 1960s onward, growing awareness of the human impact on nature fueled the rise of the environmental movement in North America, Europe, and elsewhere. It was within this context that the authors, particularly Rees, established their academic and professional focus. Important milestones of this period include the founding of Greenpeace* by a group of Canadian environmental activists in 1971, and the publication of groundbreaking reports such as the United Nations* World Commission on Environment and Development's *Our Common Future* in 1987. Popular books, such as the US ecologist Rachel Carson's* *Silent Spring* (1962), also substantially enhanced public awareness of environmental issues.

Authors' Background

Wackernagel's upbringing in Switzerland, a country with a significant share of its land dedicated to agriculture, led him to look frequently at farms (and particularly cattle-raising) to illustrate the concept of the ecological footprint. Wackernagel's PhD thesis, where this concept was initially developed, contains detailed calculations drawing on the physical laws of thermodynamics* to show that the amount of energy we can produce is dependent on the limitations of the resources we have available to generate it, echoing the conversation he had with his grandfather as an 11 year old. In an interview posted by the Global Footprint Network in 2009, Wackernagel stressed the importance of moderating our present actions in order to leave the resources needed for his son's generation, reinforcing Wackernagel's strong connection to family.[4] Following the publication of *Our Ecological Footprint*, Wackernagel has worked to implement and refine the ecological footprint model in several international contexts, and he currently serves as the president of the Global Footprint Network.[5]

On his family's farm, William Rees experienced a firsthand relationship with land, and this had a profound impact on his academic focus on human ecology,* a discipline he understands as, "the study of the interrelationships between people and their environments, as well as among individuals within an environment."[6] Professor of ecology and evolutionary biology, J. Bruce Falls* supervised Rees's PhD dissertation on the comparative ecology of bird species, and William Catton's* work on ecological overshoot* ("growth beyond an area's carrying capacity leading to die-off"[7]) served as a major influence on Rees.[8] Rees holds a doctorate in population ecology, and in 1969 he became a professor at the University of British Columbia (UBC).[9]

However, Rees's academic career had an inauspicious start because, in the early 1970s, his ideas on human carrying capacity* (the limit to the amount of human consumption the land can support) were not well received at UBC.[10] While mainstream international trade theory* at the time generally viewed a lack of regulation as broadly beneficial, he argued that global trade practices allowed developed countries to exploit the natural resources of the developing world to a devastating extent.

Despite some initial skepticism toward these ideas, the ecological footprint concept is now widely used as a policy and educational tool. Wackernagel and Rees have been recognized for their contributions to the environmental sciences, as was demonstrated when they were awarded the Kenneth E. Boulding Award and the Blue Planet Prize in 2012.[11]

NOTES

1   Mathis Wackernagel, "The Naïve Child," Dailymotion, accessed January 24, 2016, http://www.dailymotion.com/video/x3cq0r_mathis-wackernagel-the-naive-child_people.

2   Paraphrased from: Email, William Rees to Bryan Gibson, June 19, 2015.

3   M. Gismondi, "Dr William Rees: The Green Interview," *Aurora Online*, November 2000, accessed January 24, 2016, http://aurora.icaap.org/index.php/aurora/article/view/18/29.

4   Mathis Wackernagel and Susan Burns, "Global Footprint Network," June 16, 2009, accessed January 24, 2016, http://www.youtube.com/watch?v=GdtlhuciDs8.

5   "Executive Team and Senior Management," Global Footprint Network, accessed January 24, 2016, http://www.footprintnetwork.org/en/index.php/GFN/page/our_team/.

6   "Human Ecology," The Free Dictionary, accessed January 24, 2016, http://medical-dictionary.thefreedictionary.com/human+ecology.

7   Catton, William R. Jr. *Overshoot: The Ecological Basis of Revolutionary Change.* (Urbana: University of Illinois Press, 1980) 1

8   Mathis Wackernagel, "What is Ecological Overshoot?," Global Footprint Network, accessed January 24, 2016, http://www.footprintnetwork.org/en/index.php/GFN/page/video_overshoot_explained/.

9   "William Rees—Short Biography," University of British Columbia, School of Community and Regional Planning, accessed January 24, 2016, http://www.scarp.ubc.ca/people/william-rees.

10   Justin Ritchie, "Bill Rees's Last Lecture," The Tyee, February 2, 2012, accessed January 24, 2016, http://thetyee.ca/Opinion/2012/02/02/Bill-Rees-Retires/.

11   "Ecological Footprint Creators Drs. Wackernagel and Rees Win Prestigious Kenneth E. Boulding Award," Global Footprint Network, accessed January 24, 2016, http://www.footprintnetwork.org/en/index.php/newsletter/bv/dr._wackernagel_wins_prestigious_kenneth_e._boulding_award.

MODULE 2
ACADEMIC CONTEXT

KEY POINTS

- *Our Ecological Footprint* is most closely associated with the field of ecological economics*—a discipline that attempts to integrate ecological and economic research to acknowledge human dependency on nature and encourage sustainable growth.

- In the 1980s and early 1990s, the sustainable development* agenda was strongly making the case for the compatibility of economic growth and environmental sustainability* (practices intended to ensure that coming generations will be able to inherit resources).

- As part of the growing field of ecological economics, *Our Ecological Footprint* challenges mainstream economic models based on unlimited growth by showing that natural resources cannot keep up with the rates of consumption and waste production in developed countries.

The Work in its Context

Mathis Wackernagel and William Rees's *Our Ecological Footprint: Reducing Human Impact on the Earth* is part of an academic field that looks at environmental and economic concerns together. This new conceptual approach, aimed at fostering global sustainability, is used in ecological economics. By the 1990s, ecologists were investigating not only how ecosystems* function, but also how certain human activities adversely impact the available resources within particular ecosystems, the variety of life forms an ecosystem can support (a concept known as biodiversity),* and global climate change.*[1]

> 66 The Ecological Footprint is not about 'how bad things are.' It is about humanity's continuing dependence on nature and what we can do to secure Earth's capacity to support a humane existence for all in the future 99
>
> Wackernagel and Rees, *Our Ecological Footprint: Reducing Human Impact on the Earth*

At the same time, ongoing globalization*—increasing cultural, political, and economic ties across continental boundaries—forced both economists and ecologists to grapple with the particularities and consequences of international trade at a larger and more geographically integrated scale than ever before. In 1977, the US economist Herman Daly* introduced the concept of "steady-state economics,"* according to which economic production is balanced with the availability of natural resources—a potent alternative to the dominant unlimited economic growth model. Daly's ideas provided an essential reference point for the ecological footprint* model.[2]

Our Ecological Footprint offered a completely conceptual framework through its "accounting" approach to environmental impact assessment by defining the ecological footprint as the amount of land and water needed to maintain a given human population. Previously, the formula that defined the carrying capacity* for a given space was calculated by dividing a population by the land area. In a revolutionary move, Wackernagel and Rees turned this formula around, so that the land area is divided by the number of people. The ecological footprint is now the world's best-known metaphor for the human impact on the planet. This concept challenges failing development strategies that are still based on the assumption that unlimited economic growth is desirable and possible, arguing that despite human capacity to innovate new technologies, economic growth is unavoidably limited by biophysical* constraints—the limits of biological processes given the laws of physics.

The authors' work was stimulated by the increasing urgency of environmental problems, and by the stark socioeconomic imbalance associated with the overuse of natural resources. By the mid-1980s, scientists estimated that humans consumed a startling 40 percent of all the products of photosynthesis* on the planet,[3] and later studies revealed that the wealthiest 20 percent of people on Earth accounted for 80 percent of the total resource consumption.[4] Photosynthesis is the process through which plants convert sunlight, carbon dioxide, and water into metabolic energy. *Our Ecological Footprint* offers a timely and practical method to measure the ecological state of the planet and provides the general public with a clear way of understanding its own environmental impact.

Overview of the Field

The authors openly acknowledge thinkers and movements that informed their work. Earlier scholarship by the ecologists Peter Vitousek* and William Catton* attempted to quantify and analyze human impacts on the Earth, while economists such as Nicholas Georgescu-Roegen* and Herman Daly investigated important connections between natural resources, energy choices, and the global economy. Other ecological economists, such as Robert Costanza,* criticized the neoliberal* economic model, according to which markets free from governmental interference can allow unlimited economic growth, by arguing that it is not physically possible for the economy to grow infinitely, especially with a shrinking stock of natural resources. Wackernagel and Rees's seminal text grew out of these studies, and sets itself in opposition to what is called "weak sustainability,"* an approach that relies on human ingenuity and innovation to overcome the limitations of natural resources.

For decades, the United Nations* (UN) has been the central forum for debates about the environment. The UN Brundtland Commission's* 1987 report *Our Common Future*[5] sparked discussions across the international community about how to confront ecological challenges and promote sustainable development. In 1992, the UN Conference on Environment and Development (UNCED) held an "Earth Summit" in Rio de Janeiro, and established the UN Framework Convention on Climate Change (UNFCCC). This momentous summit was the first international governmental meeting dedicated entirely to environmental issues.[6] Concerns related to economic development still dominated the global political arena, however; in particular, the 1992 UN convention agreed that the acceleration of economic growth was fundamental to the eradication of poverty. The participants viewed environmental sustainability and social equity as wholly compatible with, if not explicitly achieved by, economic growth.[7] *Our Ecological Footprint* directly confronts the logic of this growth-centered perspective. The authors argue for limits to growth, outlining alternative ways to meet human *and* environmental needs while also rebalancing the vast inequalities in how resources are distributed globally.

Academic Influences

Conceptually, the authors were directly influenced by the book *The Limits to Growth* (1972) by the Club of Rome,* a think tank based in Switzerland. The book argued that a "sustainable society ... could focus on mindfully increasing quality of life rather than on mindlessly expanding material consumption and the physical capital stock."[8] The Club of Rome's acknowledgment that our planet's resources are finite challenged the prevailing view put forward by mainstream economists like Julian Simon,* who argued that for thousands of years, human ingenuity has continuously expanded human carrying capacity,* and it will continue to do so.[9] As Wackernagel noted in his PhD thesis,

"Today the debate on how to make human activities sustainable is shaped by two camps: 'The Limits to Growth' advocates and the 'Growth of Limits' advocates."[10]

Earlier attempts to construct mathematical formulas to compare resource use to resource availability laid the foundation for Wackernagel and Rees's methodology. For example, the scholars Mario Giampietro* and David Pimentel* measured the space in square meters that humans use per kilogram of fossil fuels* as a way to convert ecological impacts into a land-use area.[11] William Catton's concept of overshoot* as "habitat takeover, habitat destruction, and the drawing down of finite ancient geological resources"[12] also provided an essential building block of Wackernagel and Rees's framework.

Our Ecological Footprint follows these intellectual currents, and offers a substantial challenge to the assumption that unlimited growth is possible in a world of finite resources.

NOTES

1   R. T. Watson et al., *Climate Change 1995: Impacts, Adaptations and Mitigation of Climate Change: Scientific–Technical Analyses* (Cambridge: Cambridge University Press, 1997).

2   See Herman E. Daly, *Steady-State Economics* (Washington, DC: Island Press, 1977, 2nd edition, 1999).

3   P. M. Vitousek et al., "Human appropriation of the products of photosynthesis," *BioScience* 36, no. 6 (1986): 372.

4   "Combating Environmental Degradation," International Fund for Agricultural Development, accessed January 24, 2016, http://www.ifad.org/events/past/hunger/envir.html.

5   "Our Common Future," World Commission on Environment and Development, Document A/42/427, August 2, 1987.

6   "Conference on Environment and Development," United Nations Earth Summit, June 3–14, 1992, accessed January 24, 2016, http://www.un.org/geninfo/bp/enviro.html.

7   "Introduction," United Nations, accessed January 24, 2016, http://www.un.org/geninfo/bp/intro.html.

8   D. H. Meadows et al., *The Limits to Growth: A Report for the Club of Rome's Project on the Predicament of Mankind* 2nd edition (New York: Universe Books, 1974).

9   J. L. Simon and H. Kahn, eds., *The Resourceful Earth: A Response to "Global 2000"* (New York: Wiley-Blackwell, 1984), 45.

10 M. Wackernagel, *Ecological Footprint and Appropriated Carrying Capacity: A Tool for Planning Toward Sustainability* (PhD diss., University of British Columbia, 1994), 71.

11 D. Pimentel and M. Giampietro, "Energy Efficiency: Assessing the Interaction Between Humans and Their Environment," *Ecological Economics* 4, no. 2 (1991).

12 William R. Catton, Jr., *Overshoot: The Ecological Basis of Revolutionary Change* (Urbana: University of Illinois Press, 1980).

MODULE 3
THE PROBLEM

KEY POINTS

- *Our Ecological Footprint* has become a central framework to help identify and reduce the impact of human activity on the environment.

- Wackernagel and Rees built on earlier concepts that focused on the Earth's resource limits, such as overshoot* and carrying capacity,* to formulate the ecological footprint.* This challenged the conventional belief in unlimited economic growth.

- The ecological footprint is a useful way of visualizing the effects of overconsumption in wealthy countries and its impact on poorer ones, showing that our choices can cause both inequality and environmental damage.

Core Question

Mathis Wackernagel and William Rees's *Our Ecological Footprint: Reducing Human Impact on the Earth* set out to define the impact of human activity on the planet in order to help us reduce that impact in a meaningful way. The book deals with the behavioral, economic, and environmental challenges, and explores their links to the long-term goals of both ecological *and* economic sustainability.* As a new concept, the ecological footprint required a thorough and convincing explanation to persuade readers that adopting more sustainable practices is both necessary and possible.

Wackernagel and Rees tackle their core question by quantifying and describing the resource requirements for individuals, cities, and countries to maintain current living standards, and then estimating how much land is needed to produce the necessary amount of food

66 Beyond a certain point, the material growth of the world economy can be purchased only at the expense of depleting natural capital and undermining the life-support services upon which we all depend. 99

Wackernagel and Rees, *Our Ecological Footprint: Reducing Human Impact on the Earth*

and energy, and how much is needed to absorb our waste. Growth in both population and personal consumption has intensified land use and resulted in less land available per person.[1]

To demonstrate their concept, the authors calculated the average Canadian's ecological footprint, and then showed how this tool could be applied across different geographical and social contexts. Finally, they challenged conventional economics and international trade theory,* which depend on the misplaced assumption of infinite growth, and suggested alternatives to encourage sustainable development.*

The Participants

The authors directly built on both the concept of ecological overshoot (the use of nature's resources beyond their limits) developed by the American social scientist William Catton* in his text *Overshoot: The Ecological Basis of Revolutionary Change* (1980), and Paul Ehrlich's* work on carrying capacity (the maximum population size an area can support without a reduction in its ability to support the same species in the future), which Ehrlich developed in his *Human Carrying Capacity, Extinctions, and Nature Reserves* (1982). Rees and Wackernagel extended these scholars' observations, which made very clear the harsh reality of exploiting natural resources beyond sustainable limits.

The ecological footprint model grew out of two significant ideas from Rees and Wackernagel's earlier research: the "regional capsule"* (a term coined by Rees in the 1970s to demonstrate that cities are sustained by lands that far exceed urban boundaries) and "appropriated carrying capacity"* (which refers to the importation of ecological capacity from distant places). Moreover, Rees had already begun to use the term "ecological footprint" in 1992, when he defined it as: "the total area of land required to sustain an urban region (its 'ecological footprint') … typically at least one order of magnitude greater than that contained within municipal boundaries or the associated built-up area."[2] Rees also described "imported carrying capacity" and "exporting ecological degradation" as key components of contemporary economic arrangements, which rely on the use of food and energy produced in distant places (imported carrying capacity), and the export of waste to other parts of the world (exporting ecological degradation).

Wackernagel's dissertation elaborated and strengthened Rees's concept by placing it within a more coherent framework that addressed social and ecological concerns at the global scale. Wackernagel also established an accounting method to determine the land required for various activities, which is integral to the concept's application across a variety of global contexts.

The Contemporary Debate

The central debate in which *Our Ecological Footprint* is situated pits environmental scholars such as those of the Switzerland-based think tank the Club of Rome* and the ecologist Paul Ehrlich, who believe that available natural resources are fundamentally finite, against technology-optimists, who, like the late economist Julian Simon,* believe that human ingenuity will continue to provide innovations that enable humans to consume resources and generate waste at the same level as ever. Within this debate, *Our Ecological Footprint* clearly sides with the former camp.

The most relevant and persuasive critiques of the ecological footprint concept are that it lacks practical definitions for "unsustainable" and "sustainable" human activities, and that it ignores the positive effects of trade and the concentration of human populations in cities. Critics argue that the footprint concept is fundamentally limited because it was developed under alternative assumptions than those held by the current economic system, and consequently it struggles to represent current conditions in a meaningful way. In fact, critics claim that the ecological footprint approach's emphasis on self-sufficiency, regionalism, and localism*— that is, devotion to and promotion of the interests of a particular locality—attaches an inherently negative connotation to the global trade of goods and services.

With a growing public awareness of environmental issues, the book has fueled higher levels of interest in sustainability issues, and offers a concrete approach to measure individual and community environmental impact. Over the past two decades, governments, nongovernmental organizations (NGOs), and businesses have been able to use this model to determine the ecological footprints of, for instance, the use of various energy sources. Notably, carbon emissions (a primary contributor to global warming) are now often called carbon footprints.* However, the stubborn notion that economic well-being and ecological sustainability are mutually exclusive has kept the debate about sustainable energy solutions unresolved. As the global population grows and habits of consumption change, the ecological footprint concept has helped us understand the consequences of these developments, and has stimulated questions about whether material wealth and economic growth are the best indicators of a successful society.

Current proponents of the ecological footprint concept remain faithful to Wackernagel and Rees's model. In a key paper in the journal *Ecological Economics*, the economist Justin Kitzes* outlines a series of improvements made to the original approach. In the end, he concludes that the core message has not changed; "the world as a whole," he says, "is operating in a state of overshoot, which is continuing to increase, with residents of high-income nations demanding more productive capacity than low-income nations."[3] The concept has also been successfully tested in several case studies. For example, Chad Monfreda* copublished a study with Wackernagel, where they confirmed that the Philippines and South Korea had recently gone into ecological deficit,* consuming more than their land area produced. This suggests that ecological overshoot is expanding into developing economies,[4] and that the ecological footprint will remain an important tool to measure the impact of development on the environment.

NOTES

1   Mathis Wackernagel and William Rees, *Our Ecological Footprint: Reducing Human Impact on the Earth* (Gabriola Island, British Columbia: New Society Publishers, 1996), 79–120.

2   M. Wackernagel and W. E. Rees, *Ecological Footprints and Appropriated Carrying Capacity: Measuring the Natural Capital Requirements of the Human Economy* (Vancouver: University of British Columbia, School of Community and Regional Planning, 1992).

3   J. Kitzes et al., "A Research Agenda for Improving National Ecological Footprint Accounts," *Ecological Economics* 68, no. 7 (2009): 2003.

4   M. Wackernagel et al., "Ecological Footprint Time Series of Austria, The Philippines, and South Korea for 1961–1999: Comparing the Conventional Approach to an 'Actual Land Area' approach," *Land Use Policy* 21, no. 3 (2004): 268.

MODULE 4
THE AUTHORS' CONTRIBUTION

KEY POINTS

- *Our Ecological Footprint* argues that in order to confront the over-exploitation of natural resources, a "strong sustainability"* approach, in which the economic system is viewed as one of the subsystems of the planet's ecosystem,* is required.

- Wackernagel and Rees gave people and institutions a simple tool to measure the consumption of resources, which can provide a basis for alternative plans to meet present and future needs.

- The ecological footprint* built on the concept of human carrying capacity,* a way of measuring how much consumption, waste, and population growth a particular area can support.

Authors' Aims

In *Our Ecological Footprint: Reducing Human Impact on the Earth*, Mathis Wackernagel and William Rees outline four main aims:

- To help us understand the limits of the planet's resources and consider the needs of future generations in our daily choices.
- To illustrate the ecological footprint calculation methodology in a clear way so that local, regional, and national authorities can successfully apply it.
- To stimulate lifestyle and behavioral changes and inspire people to conduct a more environmentally conscientious existence.
- To counter the myth that unlimited economic growth is possible in a world of finite resources.

66 The first step toward reducing our ecological impact
is to recognize that the 'environmental crisis' is less
an environmental and technical problem than it is a
behavioral and social one. It can therefore be resolved
only with the help of behavioral and social solutions. 99

Wackernagel and Rees, *Our Ecological Footprint: Reducing Human Impact on the Earth*

The authors target a broad audience of the general public, policy makers, and educators. In direct opposition to those who think that environmental conservation will sabotage economic development, Wackernagel and Rees want to convince readers that their quality of life could actually improve under a reduced ecological footprint.

The overall goal is to foster real, durable change by encouraging us to reduce our impact on the environment. The model the authors develop offers citizens and organizations—from local communities to governments and businesses—a simple yet effective tool to calculate the consumption of land required for all human activities, and by extension, a way to measure sustainability.* Wackernagel and Rees seek to address the environmental consequences of pollution, climate change,* and the over-exploitation of natural resources. To do so, they have produced a clear way of measuring individual and collective environmental impacts: the ecological footprint.

Approach

Wackernagel and Rees begin with a strong assertion: humans "are not just connected with nature—we are nature."[1] The authors are convinced that if we do not change course and learn to live within nature's limits, ecological catastrophe and geopolitical chaos will result. They also stress that behavior is more important than technology, in contrast to those who place their faith in our ability to innovate technologies that will permit unlimited economic growth without endangering the environment. Although the authors propose a range of activities and policies to facilitate a more sustainable economic system, the book's central innovation is its method of establishing an account of our circumstances; calculating the ecological footprint takes the existing carrying capacity formula and turns it round to create a clear way of defining the amount of productive land required for a range of activities. This makes the ecological impact of human societies easy to understand and compare.

The book situates itself alongside ecological economics* (a field bridging ecology* and economics, extending to psychology, anthropology, archaeology, and history) and in opposition to the more traditional field of environmental economics.*

The field of environmental economics is a branch of economics that analyzes financial impacts from environmental policies such as regulatory compliance costs. It mainly focuses on the economic effects of national or local environmental policies around the world, and attempts to assign market values to ecological goods and services. In contrast, ecological economics treats the economy as one subsystem of the global ecosystem and emphasizes the preservation of natural capital:* the stock of natural assets that yields a flow of valuable goods and services into the future.

The authors define natural capital as "a stock of natural assets that is capable of producing a sustainable flow."[2] Natural capital can be renewable (fish stocks, for example), replenishable (water or solar power), or nonrenewable (fossil fuels,* for example). Ecological economists, including the authors, tend to support "strong sustainability," which holds that economic and natural capital are complementary, but not interchangeable, and humans cannot replace or sufficiently reproduce essential natural capital. By contrast, environmental economists favor "weak sustainability,"* which claims that natural capital *can* be substituted by an equivalent amount of human-made capital (such as infrastructure, labor, or financial assets).[3] Another significant contribution of *Our Ecological Footprint* is its advocacy of strong sustainability and giving the concept greater visibility in academic and policy debates.

Contribution in Context

Wackernagel and Rees fully developed the ecological footprint concept while Rees was supervising Wackernagel's PhD dissertation. In this process, Wackernagel reviewed existing work on biophysical* assessments of human needs. "Biophysical" refers to the science that applies physics to understand biological phenomena and processes. He found that the first attempts at ecological accounting* (the use of accounting and finance theories to estimate the environmental costs of economic decisions) were made in 1768 by François Quesnay,* an eighteenth-century French economist who drew a connection between environment and economy in his observation that agricultural surpluses were essential to economic growth.[4] However, much more recent studies focusing on carrying capacity, notably by Paul Ehrlich* and William Catton,* most directly inspired the authors to develop the ecological footprint concept.

Ehrlich's work on carrying capacity offered comprehensive arguments in favor of increased nature reserves to encourage the conservation of species and to avoid ecological overshoot*[5] (defined by Catton as the over-exploitation of natural resources and its consequences). These influences, together with key influential books such as the ecologist Rachel Carson's* *Silent Spring* (1962) and the Club of Rome's* *Limits to Growth* (1972), all raised the issues of over-exploitation and pollution and brought them into high-profile academic, political, and public debates. *Our Ecological Footprint* arises directly from these earlier endeavors.

Other scholars also confronted controversies about the

interconnections between the economy and ecology, but from slightly different perspectives. In 1988, the economist Salah El Serafy* suggested that a sustainable society can safely consume nonrenewable resources if equivalent natural capital is reintroduced into the economy.[6] Wackernagel argued that this approach can be modified or interpreted to emphasize the replenishment of renewable energy assets at the same rate as the nonrenewable ones, such as fossil fuel, are being used up.[7]

Increasing consumption, improved living standards, and a growing global population require an increase of the natural capital stocks, which presents a serious challenge in a world that is likely already beyond human carrying capacity. This reality makes the debates in which Wackernagel and Rees are engaged possibly the most important in all human history.

NOTES

1   Mathis Wackernagel and William Rees, *Our Ecological Footprint: Reducing Human Impact on the Earth* (Gabriola Island, British Columbia: New Society Publishers, 1996), 7.

2   M. Wackernagel and W. E. Rees, "Perceptual and Structural Barriers to Investing in Natural Capital: Economics From an Ecological Footprint Perspective," *Ecological Economics* 20, no. 1 (1997): 4.

3   L. Illge and R. Schwarze, *A Matter of Opinion: How Ecological and Neoclassical Environmental Economists Think About Sustainability and Economics* (Berlin: German Institute for Economic Research, 2006), 11; M. Wackernagel, *Ecological Footprint and Appropriated Carrying Capacity: A Tool for Planning Toward Sustainability* (PhD diss., University of British Columbia, 1994).

4   Wackernagel and Rees, *Our Ecological Footprint*, 48.

5   Paul R. Ehrlich, "Human Carrying Capacity, Extinctions, and Nature Reserves," *BioScience* 32, no. 5 (1982).

6   Salah El Serafy, "The Proper Calculation of Income from Depletable Natural Resources," in *Environmental Accounting for Sustainable Development*, eds. Yusuf J. Ahmad, et al. (Washington, DC: World Bank, 1988).

SECTION 2
IDEAS

MODULE 5
MAIN IDEAS

KEY POINTS

- *Our Ecological Footprint* presents a convincing analysis of the dangers of overconsuming the natural resources on which our lives depend, and promotes action to reduce human impact on the environment.

- The ecological footprint* is a way of measuring the ecological impact of any community, and provides a basis on which to make more sustainable choices.

- Wackernagel and Rees's arguments are logical and use clear, accessible language accompanied by numerous illustrations, allowing their insights to speak to a wide popular audience.

Key Themes

Mathis Wackernagel and William Rees's *Our Ecological Footprint: Reducing Human Impact on the Earth* shows that current rates of resource consumption are unsustainable, and provides an approach to move toward a cleaner and more fair global society in the future. The authors borrow their definition of sustainability* from the United Nations Brundtland Commission* (1980) as a matter of meeting "the needs of the present without compromising the ability of future generations to meet their own needs."[1] In practice, developing a global sustainability strategy presents serious challenges in interpretation and policy, which the authors attempt to address.

Wackernagel and Rees clearly promote "strong sustainability"* over "weak sustainability."* While both approaches acknowledge that humans are fundamentally dependent on nature, proponents of weak sustainability argue that human-made capital can serve as a

> **❝** As a result of high population densities, the rapid rise in *per capita* energy and material consumption, and the growing dependence on trade (all of which are facilitated by technology), the ecological locations of human settlements no longer coincide with their geographical locations. **❞**
>
> Wackernagel and Rees, *Our Ecological Footprint: Reducing Human Impact on the Earth*

substitute for diminished natural capital* such as food, water, and fuel; whereas advocates for strong sustainability argue that natural stocks and human-made capital are not interchangeable.

Ecological deficit* is one of the major dangers the authors identify; the term refers to the condition experienced by communities living beyond their local ecological means. One clear illustration of this phenomenon is that many cities and wealthy communities require more resources to maintain their lifestyle and consumption rates than they produce. This leads to the exploitation of resources in distant, typically poorer, regions across the globe.

This work is grounded in a simple fact: our continued existence is dependent on the conservation of natural systems. The idea of carrying capacity,* defined as the maximum human population that can be sustained in a given area, marks the limits of nature's ability to support human activity. Wackernagel and Rees draw directly on this concept to formulate their central concept—the ecological footprint—"the land (and water) area that would be required to support a defined human population and material standard indefinitely."[2] The ecological footprint is a tool that clearly quantifies the ecological impact of any community, providing a practical way of addressing the urgent issues facing present and future societies.

Exploring the Ideas

Wackernagel and Rees point out different interpretations of what sustainability and sustainable development* actually mean. Some people focus on the ecological and social aspects of sustainability, while others focus on an environmentally sensitive version of economic growth.[3] However, the authors venture a simple definition; for them, sustainability is a question of "living in material comfort and peacefully with each other within the means of nature."[4] Despite this simplicity, the authors caution that, without a general agreement on its policy implications, progress toward constructing a more sustainable future will be slow.

They elaborate their discussion on sustainability by distinguishing between its weak and strong variants, and presenting their case for a strong sustainability approach. According to the authors, many economists favor weak sustainability, which sees a balance between human-made capital and natural capital; indeed, these economists consider them interchangeable. Under weak sustainability, the income generated by a factory can be seen as a substitute for the lost income-potential of a cleared forest.[5] Strong sustainability, on the other hand, recognizes "the unaccounted ecological services ... performed by many forms of natural capital and the considerable risks associated with their irreversible loss."[6] In brief, strong sustainability requires the maintenance of natural capital independent of its human-made counterparts. Wackernagel and Rees note the "ecological irrelevance" of weak sustainability, claiming it fails to account for the reliance of rich countries on the natural capital of poorer ones.

This global exchange of natural capital—which tends to flow from poorer to richer nations—is embodied by the concept of ecological deficit. The authors observe that "the ecological locations of human settlements no longer coincide with their geographic locations," and show the extent to which modern cities are dependent on a vast and increasing "global hinterland of ecologically productive landscapes."[7] This critical analysis of wealthier regions as living beyond the carrying capacity of their local natural capital exposes the still relatively hidden truth that comforts in the developed world are enjoyed at the ecological expense of poorer communities.[8]

Carrying capacity is a standard analytical measure in ecology* with a long intellectual history, and the authors cite it as a direct inspiration for their ecological footprint concept. Traditionally, carrying capacity has been used to determine the "maximum population of a given species that can be supported indefinitely in a specified habitat."[9] Wackernagel and Rees elaborate this concept, apply it to human populations, and invert the formula. The ecological footprint "measures land area required per person (or population), rather than population per unit area."[10]

The ecological footprint principle is simple and clear, measuring "the area of ecologically productive land (and water) … that would be required on a continual basis a) to provide all the energy/material resources consumed, and b) to absorb all the wastes discharged."[11] The authors proceed to develop a framework to calculate footprints based on several criteria. At its core, the ecological footprint is an accounting tool that allows anyone to estimate the extent of global overshoot* (the amount by which humanity's footprint exceeds global carrying capacity) and the ecological deficit of households, companies, regions, or countries.[12]

Language and Expression

The ideas expressed in *Our Ecological Footprint* are presented in a logical series: the authors define the core concepts; produce empirical observations that demonstrate resource overuse; introduce and illustrate their ecological footprint concept and methodology; and, finally, propose some specific steps to move economy and society in a more sustainable direction.

These ideas are expressed in well-articulated arguments that use clear and accessible language and contain many illustrations. Cities are graphically represented as a human foot to make a straightforward connection between the ecological footprint metaphor and the impact of urban life on local and distant environments. The authors provide more intricate detail when needed, especially to describe the calculation methodology and potential applications of ecological footprint analysis. They also challenge their readers' imaginations, using lively imagery—a city enclosed in a dome, for example, to show that urban life is not actually closed off from nature, but wholly dependent on it.

It is clear that the authors hope to convey both the urgency of the challenge humanity faces and a sense that solutions are possible through sensible evaluations of the problem and collective action to make meaningful change. The tone and organization are straightforward and rousing, reflecting the authors' intention to inspire a wide, global readership to join the strong sustainability mission.

NOTES

1 Mathis Wackernagel and William Rees, *Our Ecological Footprint: Reducing Human Impact on the Earth* (Gabriola Island, British Columbia: New Society Publishers, 1996), 33.

2 Wackernagel and Rees, *Our Ecological Footprint*, 158.

3 Wackernagel and Rees, *Our Ecological Footprint*, 32–3.

4 Wackernagel and Rees, *Our Ecological Footprint*, 32.

5 Wackernagel and Rees, *Our Ecological Footprint*, 37.

6 Wackernagel and Rees, *Our Ecological Footprint*, 37.

7 Wackernagel and Rees, *Our Ecological Footprint*, 29.

8 For examples, see Wackernagel and Rees, *Our Ecological Footprint*, 96–100.

9 Wackernagel and Rees, *Our Ecological Footprint*, 48–9.

10 Wackernagel and Rees, *Our Ecological Footprint*, 51.

11 Wackernagel and Rees, *Our Ecological Footprint*, 51–2.

12 Wackernagel and Rees, *Our Ecological Footprint*, 55.

MODULE 6
SECONDARY IDEAS

KEY POINTS

- *Our Ecological Footprint*'s secondary ideas include entropy* (the physical law that an isolated system will degenerate into a more disordered state), natural capital* and natural income,* questions about competing energy resources, and the ecological footprint's* relationship to the marine environment.

- These are important issues that help the authors state their central arguments, produce avenues for further research, and offer a refinement of the ecological footprint concept.

- Of these secondary themes, the debate between energy resources has become the most important due to intense public debates about how to address climate change* and reduce carbon footprints* (the total amount of carbon compounds emitted in the course of human activities).

Other Ideas

The notion of entropy from the second law of thermodynamics* provides a solid base in physics for Mathis Wackernagel and William Rees's ideas in *Our Ecological Footprint: Reducing Human Impact on the Earth*. The second law of thermodynamics (the physical laws dealing with the conversions between various forms of energy, including heat) is the "entropy law," according to which the entropy—or disorder—of a system always increases. A system that eventually uses up all the energy available to it will run down.[1]

The authors describe the human economy as a dynamic, open, and growing subsystem of a closed nongrowing ecosphere* (which is, in

> 66 Beyond a certain point, the continuous growth of the economy ... can be purchased only at the expense of increasing disorder (entropy) in the ecosphere. 99
>
> Wackernagel and Rees, *Our Ecological Footprint: Reducing Human Impact on the Earth*

essence, an area that can support life, and the relationship between living organisms and their environment).

The ecosphere provides the economic subsystem with energy and absorbs its waste; however, its energy stocks and absorption capacity are still limited and subject to the second law of thermodynamics, meaning that the energy will eventually be lost as entropy, for example in the form of heat.

Wackernagel and Rees also deployed the idea of natural capital to think about the best ways to use resources on a long-term basis. Natural capital is defined as "any stock of natural assets that yields a flow of valuable goods and services into the future."[2] Examples include forests or fish stocks which may provide useful goods sustainably year after year; in this case, the stock is natural capital, and the sustainable harvest is called natural income. If we consume more than the income from our natural capital, we will diminish our natural wealth and reduce the capital available to future generations. This metaphor helps readers to understand how our use of natural resources now has serious consequences for the future.

Importantly, the authors touch on a larger debate about competing energy sources—renewable energy sources, fossil fuels* (oil, gas, and coal, largely) and nuclear. To weigh up the best options, they include the concept of "energy land"* as a part of their ecological footprint calculation.* They conclude that renewable sources, such as hydroelectric and solar energy, are much more productive in terms of energy units per hectare of land, and therefore contribute less to the ecological footprint than fossil fuels. In their view, this clearly shows that renewable energy production should be expanded because it is more sustainable and requires less "energy land."

Though some ecologists are attracted to nuclear energy as a low-pollution source, Wackernagel and Rees dismiss nuclear energy as undesirable because of the unresolved challenge of radioactive waste disposal.

Finally, the book addresses the human impact on the marine environment. Originally, this was excluded from the ecological footprint calculations because fish stocks provide only a small proportion of overall nutritional energy, and humans are less able to consciously manipulate the seascape to the same extent as land. However, fish stocks are dramatically overharvested, and an initial estimation of the footprint of fisheries worldwide equals 0.51 hectares per capita,[3] a significant proportion. The scale and urgency of the problems with our seas make this an issue the authors could not ignore.

Exploring the Ideas

These secondary ideas about entropy, natural capital and income, competing energy sources, and the ecological footprint's relationship to the marine environment are presented concisely and, for the most part, set out in separate text boxes throughout the book. But this should not suggest that these points are inconsequential. The authors' discussion of the second law of thermodynamics reveals a major shortcoming of neoclassical economics*—namely, the failure of these policies to situate the economy in its natural (environmental) context. Entropy tells us that unsustainable economic growth comes "at the expense of increasing disorder at high levels in the systems hierarchy" (in other words, the ecosphere).[4] The authors cite this as a reality check to advocate for an alternative economic theory that accounts for the unavoidable laws of entropy.

The concepts of natural capital and natural income play a related role in the book. The inclusion of natural capital marks a significant revision to traditional economic models, and helps set up the debate the authors construct between "weak" and "strong" sustainability* approaches. Strong sustainability requires that we "conserve or enhance our natural capital stocks," whereas weak sustainability* argues that "losses of natural capital [are] acceptable if compensated through the substitution of an equivalent amount or value of human-made capital."[5] Both entropy and natural capital/income suggest the need for an alternative economic model that incorporates ecological and thermodynamic* considerations; for Wackernagel and Rees, ecological economics* allows us to think about ecosystems* and economies together, as they are in reality. This is a field of inquiry bridging ecology* and economics, and which commonly extends to psychology, anthropology, archaeology, and history.

The debate between competing energy sources factors directly into the method used to define and measure the ecological footprint. To determine the amount of "energy land" to be incorporated into footprint analysis, the authors had to construct a framework to compare the productivity of various energy sources, showing that "the higher the productivity, the smaller the footprint."[6] In this sense, the debate about the best energy sources is incidental to the construction of the footprint; however, given the public and political debates about energy policies and the production of carbon dioxide and other greenhouse gases responsible for man-made climate change, the authors' energy productivity calculations—which promote renewables—make them an important voice in this ongoing debate.

Our Ecological Footprint provides only a "generalized first approximation" of the human impact on marine fisheries,[7] and does not address why this issue is not directly addressed in their analysis. However, the ecological footprint on the sea remains important as a secondary theme because it suggests a pathway for future research, as well as a refinement and expansion of the footprint concept.

Overlooked

Given the widespread success of *Our Ecological Footprint*, no major parts have been neglected by contemporary thinkers. Nevertheless, there are areas that could have received more attention and have not been discussed in depth by subsequent works because of various commentators' points of view. For example, Wackernagel and Rees strongly challenged the validity of assessing sustainability* and resource scarcity by assigning a monetary value to services provided by, or damage done to, the environment—an approach favored by environmental economists.*[8] While it may seem strange that this critique was not picked up and defended by environmental economists, this apparent oversight may be intentional to avoid a crucial disagreement between schools of thought with broadly the same aims.

Similarly, the authors' discussion about efficiency and resource use has not provoked wide debate largely because of ideological barriers. In the section of the book with the subtitle *Will efficiency gains save resources?*, the authors argue that technological-optimism is naïve; if the world's population and living standards continue to grow, the total consumption increase will likely outweigh any technological gains in efficiency.[9] Pervasive technological enthusiasm and a long history of associations between conservation and efficiency gains has, however, left this aspect of their work under-discussed.[10]

Overall, therefore, the book has been thoroughly discussed in contemporary scholarship and in public debate, although as the examples above show, a diversity of opinion about the best way forward toward a sustainable future sometimes prevents total consensus on these issues.

NOTES

1   Mathis Wackernagel and William Rees, *Our Ecological Footprint: Reducing Human Impact on the Earth* (Gabriola Island, British Columbia: New Society Publishers, 1996), 43.

2   Wackernagel and Rees, *Our Ecological Footprint*, 35.

3   Wackernagel and Rees, *Our Ecological Footprint*, 65.

4   Wackernagel and Rees, *Our Ecological Footprint*, 43.

5   Wackernagel and Rees, *Our Ecological Footprint*, 36.

6   Wackernagel and Rees, *Our Ecological Footprint*, 69.

7   Wackernagel and Rees, *Our Ecological Footprint*, 65.

8   Wackernagel and Rees, *Our Ecological Footprint*, 40–8.

9   Wackernagel and Rees, *Our Ecological Footprint*, 128–9.

10 For example, see Samuel P. Hays, *Conservation and the Gospel of Efficiency: The Progressive Conservation Movement, 1890-1920* (Cambridge, MA: Harvard University Press, 1959).

MODULE 7
ACHIEVEMENT

KEY POINTS

- *Our Ecological Footprint* successfully developed and communicated both a useful ecological accounting* tool and a vision of sustainability* and the equitable use of natural resources across cultures and countries.

- Wackernagel and Rees's idea of the ecological footprint* has been widely adopted by policy makers and the general public alike, and has become a popular way of thinking and talking about the human impact on the environment.

- While ideological barriers—especially from conventional economic theory—restricted the reception of the authors' ideas, the most serious limitations of the ecological footprint are its oversimplifications and a perspective that does not easily account for changes over time.

Assessing the Argument

In *Our Ecological Footprint: Reducing Human Impact on the Earth*, Mathis Wackernagel and William Rees introduce their central ideas and methodology, concisely providing readers with a way to use the tools they present. The authors challenge neoclassical economics,* an approach that equates economic health with constant growth, and offer a set of recommendations on how to avoid the use of natural resources beyond their limits: ecological overshoot.* It was essential to the authors that their ideas reach ordinary people, so their approach might be considered too straightforward for experts looking for greater methodological and conceptual detail. For specialists or those wishing to go further into these issues, additional resources are available in the book's reference material. Overall, the

> ❝ In summary, monetary approaches are blind to the requirements for ecological sustainability because they do not adequately reflect biophysical scarcity, social equity, ecological community, incommensurability, structural and functional integrity, temporal discontinuity, and complex system behavior. ❞
>
> Wackernagel and Rees, *Our Ecological Footprint: Reducing Human Impact on the Earth*

authors succeeded in communicating to a general audience both an innovative approach to sustainability and a useful way of measuring environmental impact.

Still, it can be argued that the authors could have discussed more of the complex problems confronted by governments during international negotiations, such as the 1992 Rio Earth Summit, and expanded their criticism to specific claims made by international environmental conventions. In their discussion of the Rio Earth Summit, they did not directly critique the summit's main outcome, the Rio Declaration, for its continued focus on economic growth.[1] Although the book constantly reminds readers that unlimited growth and sustainability are incompatible objectives, the authors could have been firmer in their critique of this contradiction through more specific case studies of international environmental agreements. These choices might show that by avoiding these topics, Wackernagel and Rees prioritized reaching a wide audience over investigating certain issues in greater depth.[2]

Achievement in Context

Our Ecological Footprint is still relevant and scholars continue to discuss the text's usefulness and limitations. Some subsequent publications sought to diminish the value of the ecological footprint concept, while others have enhanced it by proposing modifications and improvements. The agricultural professor Nathan Fiala* highlighted how Wackernagel and Rees's approach measures inequality of resource distribution but not sustainability, which, in his view, means that it is unable to discern whether land use is sustainable or unsustainable.[3] Furthermore, Fiala argued that their calculations of the planet's biocapacity*—that is, how much life the planet can sustain—are based on average consumption figures, and don't take into consideration the ways that future technological improvements could help increase biocapacity.

The economist Fabio Grazi* extended this criticism, applying his own framework—called spatial welfare analysis*—to the same regions studied by Wackernagel and Rees, and found opposite results. Grazi concluded that the ecological footprint cannot detect sustainable development* that also increases people's welfare, a central part of most development strategies.[4] By contrast, the political scientist Chad Monfreda* developed an enhanced ecological footprint tool that more clearly distinguishes between primary production (such as the creation of new organic matter by plants through the natural process of photosynthesis)* and secondary production (which includes the creation of new organic material by animals like plankton).[5]

Since its publication in 1996, more elaborate analyses conducted following the original ecological footprint approach have added value by introducing more detailed data and calculations. But there are still deep-rooted divisions among thinkers in this field. According to the Dutch economists G. Cornelis van Kooten* and Erwin Bulte,* the ecological footprint "may prove anathema to those who are concerned about the environment and sustainable development" due to its lack of sound methodology and its vague assumptions.[6] For others, like Grazi, its shortcomings provide a useful starting point to construct more reliable tools to estimate human impacts on the environment.

Some scholars saw the ecological footprint "as being useful in a different way for policy-making."[7] In particular, the British sustainability scholar Ian Moffatt* suggested that the ecological footprint could be combined with input–output analysis* (a mathematical tool used in economics to analyze how goods and services flow between sectors) or natural resource accounting to develop a dynamic model that could be integrated with Geographical Information Systems* (or GIS—a data management system to store and manipulate geographical information.)[8] The resources scholar David Rapport* suggested that the ecological footprint and other approaches concerned with the health of an ecosystem* could be used together to assess both natural resource exploitation and ecosystem degradation.[9] This last argument is important when we consider land degradation and unsustainable land uses that reduce the health of ecosystems, but are not easily accounted for in the original ecological footprint methodology.

Limitations

In *Our Ecological Footprint*, Wackernagel and Rees produced a concept and a methodology meant to be universally applicable. However, because ecological footprinting is an approach that takes a snapshot of a community's resource use at a particular moment, historical ecological data is needed to allow comparisons across space and time. For instance, large countries such as Canada or Australia have much higher biocapacity levels—the capacity to produce an ongoing supply of renewable resources and to absorb waste—because of their abundance of ecologically productive land, whereas smaller countries, such as the Netherlands or Switzerland, have an ecological deficit* (that is, the land cannot support the demands made of it by the population) simply because they have less productive land. Therefore, the per-capita consumption of an Australian citizen is (logically) likely to be higher than that of a

Dutch person. The authors appear to suggest that each nation or community should make do with their own resources, and that long-distance trade is intrinsically negative. This view portrays trade and the global capitalistic* system negatively, and means that the authors reject out of hand potential solutions that would operate within the mainstream economic system.

A second limitation is that the authors tend to oversimplify resource use and waste disposal by trying to translate all human activities into a single land use measurement. While Wackernagel and Rees confess that their deliberate simplifications are open to criticism,[10] they lay out their rationale for keeping footprint calculations* simple and transparent. They make "operational" simplifications in order to make the model widely useable, and claim that ecological footprint analyses "need not include all consumption items, waste categories and ecosphere* functions to have diagnostic value."[11]

NOTES

1   United Nations Conference on Environment and Development, "The Rio Declaration on Environment and Development," United Nations Environment Programme, accessed January 24, 2016, http://www.unep.org/documents.multilingual/default.asp?documentid=78&articleid=1163.

2   W. E. Rees, "Eco-Footprint Analysis: Merits and Brickbats," *Ecological Economics* 32, no. 3 (2000): 371–4.

3   N. Fiala, "Measuring Sustainability: Why the Ecological Footprint is Bad Economics and Bad Environmental Science," *Ecological Economics* 67, no. 4 (2008): 520.

4   F. Grazi et al., "Spatial Welfare Economics versus Ecological Footprint: Modeling Agglomeration, Externalities and Trade," *Environmental and Resource Economics* 38, no. 1 (2007): 151–2.

5   C. Monfreda et al., "Establishing National Natural Capital Accounts Based on Detailed Ecological Footprint and Biological Capacity Assessments," *Land Use Policy* 21, no. 3 (2004): 231.

6   G. C. Kooten and E. H. Bulte, "The Ecological Footprint: Useful Science or Politics?" *Ecological Economics* 32, no. 3 (2000): 385–9.

7   R. Costanza, "The Dynamics of the Ecological Footprint Concept," *Ecological Economics* 32, no. 3 (2000).

8   I. Moffatt, "Ecological Footprints and Sustainable Development," *Ecological Economics* 32, no. 3 (2000): 361.

9   D. J. Rapport, "Ecological Footprints and Ecosystem Health: Complementary Approaches to a Sustainable Future," *Ecological Economics* 32, no. 3 (2000): 369.

10 Mathis Wackernagel and William Rees, *Our Ecological Footprint: Reducing Human Impact on the Earth* (Gabriola Island, British Columbia: New Society Publishers, 1996), 62–3.

11 Wackernagel and Rees, *Our Ecological Footprint*, 63.

MODULE 8
PLACE IN THE AUTHORS' WORK

KEY POINTS

- Wackernagel and Rees's academic careers have largely focused on developing the ecological footprint* model and other ideas drawn from human ecology* (the study of the relationships between humans and nature), and ecological economics.* Following the success of *Our Ecological Footprint*, the authors have continued to spread and apply the ecological footprint through their nongovernmental organizations (NGOs).

- For William Rees, the ecological footprint represents a high point of his career. For Mathis Wackernagel, it marked the beginning, and the basis for his later work.

- *Our Ecological Footprint* is central to both authors' reputations as well as the focal point of their scholarship and advocacy work.

Positioning

Mathis Wackernagel and William Rees's *Our Ecological Footprint: Reducing Human Impact on the Earth* merges the work of a senior intellectual, Rees, with that of a young scholar, Wackernagel, who was once one of his PhD students. Having supervised Wackernagel's doctorate, Rees took him under his wing and helped him further develop what he viewed as a groundbreaking model for determining the impact humans were having on their environment. In the early 1990s, the authors began to work on early versions of the ecological footprint concept and spent a considerable amount of time earning academic credibility and popular support through conference presentations and other speaking engagements.

❝ How secure are ecological surpluses upon which developed regions currently depend? What does this dependency mean for the potential intensification of local and global resource conflicts? What sorts of new international agreements are needed to formalize stable relationships among interdependent regions? **❞**

Wackernagel and Rees, *Our Ecological Footprint: Reducing Human Impact on the Earth*

When *Our Ecological Footprint* was published, Rees was already an established scholar at the University of British Columbia. After publishing the text, he went on to found the One Earth initiative,*[1] a Canadian research and advocacy group promoting long-term sustainability.* For Wackernagel, the importance of this book to his career is even more considerable. It is quite rare for a young scholar to produce a seminal text for their first publication. This effectively launched his career and he went on to found the successful Global Footprint Network* (GFN).[2] In short, this book presented both authors with opportunities to broaden their horizons beyond the academic world and represents a significant career milestone for each of them.

Integration

The authors' books and publications showcase original ideas based on sound research that continues to influence fellow academics, policymakers, and public opinion. At the moment, however, the ecological footprint concept is not as publically prominent as debates around climate change,* which tend to focus on strategies to limit global warming, while issues such as pollution and species extinction have to some extent been sidelined. In Rees's words, "catastrophe forces people to action much more than new information does."[3]

Following his argument, without a tremendous wake-up call for humanity, such as the melting of the Greenland ice-sheet, the valuable efforts by Wackernagel and Rees (and many of their colleagues) will not be enough to change our relationship with the planet. Although the extent of the authors' achievement is ongoing and far-reaching, it remains functionally limited in the face of public and governmental inaction.

Our Ecological Footprint laid the foundation for Rees's subsequent insights about human evolution. For Rees, "evolution has not provided us with inhibitions against … destroying our earthly habitat(s)."[4] While the book focused mainly on human populations, the authors' later work took a broader view and also examined the impact of environmental decline on other species. Wackernagel and Rees's later publications also engage with the realities of contemporary economics to a greater extent, eventually incorporating arguments that concede that, "high eco-efficiency and competitiveness [are] not mutually exclusive."[5] Both authors remain engaged in high-profile discussions about the future of the world and economy. Their ideas continued to be developed in the late 1990s and early 2000s to better elaborate the tensions and opportunities presented by globalization* and sustainability. Their work is diverse, but coherently placed in the domains of human ecology and ecological economics (a field that bridges ecology* and economics, routinely drawing on those of psychology, anthropology, archaeology, and history).

Significance

Our Ecological Footprint remains the focal point of the authors' reputations. At the time of the book's publication, Rees was an established scholar with a strong record of academic production; *Our Ecological Footprint* strengthened his reputation in the field and increased his scholarly and public profile. Wackernagel's reputation is founded on the ecological footprint concept. The book and concept emerged from his PhD dissertation and marks his entrance into academic discourse. His subsequent work—mainly with his NGO, the Global Footprint Network*—and academic publications are largely focused on the application, defense, and extension of the ecological footprint.

Our Ecological Footprint has inspired the development of an academic community focused on advancing the authors' concepts and agenda. For example, Nicky Chambers* and Craig Simmons,* cofounders of "Best Foot Forward"—the UK's first environmental consultancy that works entirely on ecological footprinting—have advanced the practical application of the ecological footprint. With Wackernagel, they coauthored *Sharing Nature's Interest* (2000), a guide on how to make human societies live on natural income (whatever can be replenished by the Earth's regenerative capacity) rather than natural capital* (the stock of natural assets that yields a flow of valuable goods and services into the future). The book has also found support among established scholars working on corresponding ideas. For example, the sustainability scholar Robert Costanza* supported *Our Ecological Footprint*'s strong sustainability* framework, holding up the ecological footprint model as a "useful provisional indicator of sustainability at the global scale," particularly because the model "assumes that technology will not save us."[6]

The publication of *Our Ecological Footprint* was a "game changer" in the authors' careers, introducing a model with a deep and enduring impact in science, policy, and society at large. As ecology as a scientific discipline grows and matures, and with the increased public awareness of environmental issues, *Our Ecological Footprint* has provided a timely contribution to positive change at a number of social levels. Ecological footprint and carbon footprint calculators* (methods of calculating how much carbon is required in the performance of human activities) derived from the Global Footprint Network, have been implemented and published online by many environmental NGOs, governments, and businesses.[7] The ecological footprint is being adopted by national governments, and the European Union, as an environmental indicator on which to base their choices and policies.[8]

These successes demonstrate both the significance of the work in the authors' ongoing careers, and the public and academic profile of the idea of the ecological footprint.

NOTES

1   "Home," One Earth, accessed January 24, 2016, http://oneearthweb.org/.

2   "Home," Global Footprint Network, accessed January 24, 2016, http://www.footprintnetwork.org/en/index.php/GFN/.

3   William Rees, "How to Convince People to Face Reality," Citizen Action Monitor, September 22, 2011, accessed January 24, 2016, http://citizenactionmonitor.wordpress.com/2011/09/22/william-rees-we-need-a-mass-social-movement-but-time-is-not-on-our-side/.

4   W. E. Rees, "Is Humanity Fatally Successful?" *Journal of Business Administration and Policy Analysis* (2004): 67.

5   A. Sturm et al., *The Winners and Losers in Global Competition: Why Eco-efficiency Reinforces Competitiveness—A Study of 44 Nations* (West Lafayette, IN: Purdue University Press, 2003).

6   R. Costanza, "The Dynamics of the Ecological Footprint Concept," *Ecological Economics* 32, no. 3 (2000): 342.

7   "Personal Footprint," Global Footprint Network, accessed January 24, 2016, http://www.footprintnetwork.org/pt/index.php/GFN/page/personal_footprint/.

8   "Case Stories," Global Footprint Network, accessed January 24, 2016, http://www.footprintnetwork.org/pt/index.php/GFN/page/case_stories/; "Europe 2007 Gross Domestic Product and Ecological Footprint," World Wildlife Fund, accessed January 24, 2016, http://www.wwf.se/source.php/1149816/Europe%202007%20-%20GDP%20and%20Ecological%20Footprint.pdf.

SECTION 3
IMPACT

MODULE 9
THE FIRST RESPONSES

KEY POINTS

- *Our Ecological Footprint* received several important criticisms, many of which have actually been beneficial, helping the authors and other thinkers refine the ecological footprint* model and create new ones.

- Wackernagel and Rees responded to these criticisms by acknowledging the ecological footprint's limitations—but declaring also that it was intended to provide a big-picture analysis of various human activities' impacts on the environment.

- Although fundamental disagreements about unlimited economic growth have divided its audience, this debate has not undermined the value of ecological footprinting as an important tool.

Criticism

Mathis Wackernagel and William Rees's *Our Ecological Footprint: Reducing Human Impact on the Earth* received a variety of reactions from the scientific community and policy makers. In general, critiques of the book focused on the simplicity of its calculations, its stance against international trade, and its analysis of land use. The sustainability* consultant Roger Levett* points out that although the calculation of the ecological footprint helps to *frame* questions about the ethics, politics, and management of the transition to a more sustainable global society, it cannot *answer* them due to methodological weaknesses and implementation problems.[1]

Another important line of criticism was put forward by the Dutch environmental economist* Jeroen van den Bergh* and the

> 66 The strength of the ecological footprint analysis is its ability to communicate simply and graphically the general nature and magnitude of the biophysical 'connectedness' between humankind and the ecosphere. 99
>
> Wackernagel and Rees, *Our Ecological Footprint: Reducing Human Impact on the Earth*

marine biologist Heroen Verbruggen,* who noted four main problems with the text:

- There are no clear policy objectives contained in the model.
- It does not provide definitions of sustainable and unsustainable land use.
- The calculation of ecological footprints at the national level are so large that they become abstract and less meaningful, and the model does not recognize the positive impact of population density of cities, providing for more people in less space.
- The approach neglects possible positive effects of trade, which tends to spread the environmental burden rather than keeping it concentrated in one place.[2]

As the debate evolves, the focus of Wackernagel and Rees's critics has shifted away from specific technical problems with the ecological footprint model to proposals about how to improve the model itself. In particular, the environmental economists Jason Venetoulis* and John Talberth* developed a new methodology, which includes the whole Earth's surface in their biocapacity* estimates and considers the impact of other species as well. They concluded that humanity's global footprint is much greater than Wackernagel and Rees first suggested.[3]

Responses

In response to these critiques, Wackernagel and Rees acknowledged that the aim of the ecological footprint was to provide a big-picture analysis and to contextualize the various competing human uses of the biosphere* (the part of the Earth and its atmosphere capable of supporting life). Using the example of freshwater use and contamination, they note that these values "can be only estimated through proxy calculations."[4] Importantly, the authors realized that ecological footprint analysis does not measure quality of life and, accordingly, they suggested that social indicators should also be evaluated to design more robust sustainability strategies.

Wackernagel clarified that calculating the impact of pollution may be difficult or even impossible, and that available data sources, such as United Nations* databases, offered incomplete information. Wackernagel and his associate Judith Silverstein* responded to van den Bergh and Verbruggen's critiques by arguing that distributing environmental costs to less ecologically vulnerable areas is based on a false assumption of constant demand, and could lead to increased resource consumption and more pervasive environmental impacts.[5]

The critics of the ecological footprint inspired the authors to improve their methodology and explain their ideas within the framework and language of competing economic theories. However, since their theory and vision of the world fundamentally challenges the dominant assumptions of neoclassical economics,* which influence human activity the world over, notable differences persist and the debate continues.

Conflict and Consensus

Most scholars agree that the ecological footprint is a valid analytical and descriptive tool and have applauded Wackernagel and Rees for their contribution. But many critics did not share the authors' view that the ecological footprint can be used as a robust policy tool; even allies in the field of ecological economics* seem to be cautious in this regard.

While Wackernagel and Rees did not significantly revise their original ideas in subsequent editions of *Our Ecological Footprint*, they have offered defenses and refinements in other publications and in collaborations with colleagues. Notably, in an article of 2000, Wackernagel and Silverstein specifically countered van den Bergh and Verbruggen's critique that national boundaries do not have an environmental meaning by stating that policy decisions taken inside national boundaries invariably affect the ecosystems* in that nation.[6] Other significant critiques point out that *Our Ecological Footprint* presents international trade as inherently damaging and takes the value of self-sufficiency (resource independence) for granted; it could be noted that Wackernagel and Rees have not directly addressed or defended these points.

Regardless, dialogue and critical analysis of the work has continued over the last decade, and scholars have enhanced the ecological footprint's potential to serve as a robust planning instrument.

Critiques of *Our Ecological Footprint* have actually been beneficial, prompting thinkers and policymakers to develop more comprehensive and resilient tools for sustainability planning.* The ecological footprint helped open up discussions between civil society, policymakers, and the media by providing a simple reference number that approximately represents resource consumption and environmental impact. This has been revolutionary in terms of raising awareness of individual and collective responsibility and the ecological implications of various human activities. In fact, the tool allows us to estimate how much land we use directly (say, by growing crops for food) and indirectly (in the case of forests absorbing carbon emissions, for example),[7] as well as exposing the dependency of wealthy countries on land in less developed, but resource-rich, countries.[8]

The ecological footprint offers an indication of how a consumption-based lifestyle does not just damage the Earth but also drives global inequality. As debates about global warming intensify, the public conversation is centered on energy and associated carbon emissions and has become less focused on overall resource consumption. Although the public debate may temporarily shift away from the broad resource-based focus of *Our Ecological Footprint*, the concept Wackernagel and Rees produced remains an important tool to quantify and visualize the ecological impacts of our existence on this planet.

NOTES

1   R. Levett, "Footprinting: A Great Step Forward, But Tread Carefully—A Response to Mathis Wackernagel," *Local Environment: The International Journal of Justice and Sustainability* 3, no. 1 (1998): 73–4.

2   J. C. J. M. van den Bergh and H. Verbruggen, "Spatial Sustainability, Trade and Indicators: An Evaluation of The 'Ecological Footprint'," *Ecological Economics* 29, no. 1 (1999).

3   J. Venetoulis and J. Talberth, "Refining The Ecological Footprint," *Environment, Development and Sustainability* 10, no. 4 (2008): 441.

4   M. Wackernagel et al., "National Natural Capital Accounting with The Ecological Footprint Concept," *Ecological Economics* 29, no. 3 (1999): 387.

5   Van den Bergh and Verbruggen, "Spatial Sustainability," 441.

6   M. Wackernagel and J. Silverstein, "Big Things First: Focusing on the Scale Imperative with the Ecological Footprint," *Ecological Economics* 32, no. 3 (2000): 393.

7   Mathis Wackernagel and William Rees, *Our Ecological Footprint: Reducing Human Impact on the Earth* (Gabriola Island, British Columbia: New Society Publishers, 1996), 88–91.

8   S. Bond, *Ecological Footprints: A Guide for Local Authorities* (Godalming: World Wildlife Fund UK, 2002), 4.

MODULE 10
THE EVOLVING DEBATE

KEY POINTS

- *Our Ecological Footprint*'s most important contributions to later scholars is its critique of the failure of dominant economic ideas to consider both human dependence on nature and the evolution of alternative sustainable development* strategies.

- The book highlighted the differences between two competing schools of thought (ecological economics* and environmental economics),* their different approaches to sustainability* (strong* and weak)*, and the appropriateness of assigning monetary value to nature.

- *Our Ecological Footprint* has significantly influenced local communities, politicians, environmental organizations, and the media.

Uses and Problems

Mathis Wackernagel and William Rees's *Our Ecological Footprint: Reducing Human Impact on the Earth* has had a revolutionary influence in the fields of ecology* and economics, and the social sciences. Since its publication in 1996, the book has been a touchstone in debates about sustainability and alternative development strategies. In particular, the *Journal of Ecological Economics* has published a wide range of articles that discuss the ecological footprint,* including a series of articles in 2000 that emerged from a forum on the concept of the ecological footprint. Pivotal discussions between environmental economists (who focus on the analysis of the financial impacts of environmental policies) and ecological economists* (whose inquiry draws on fields such as ecology, economics, anthropology, and

> **"** In some cases, once nature has been overexploited, no amount of manufactured goods will compensate for the loss of natural capital. **"**
>
> Wackernagel and Rees, *Our Ecological Footprint: Reducing Human Impact on the Earth*

archaeology) arose out of this series about how society can become more sustainable and self-sufficient, and to what extent trade systems should be regulated to safeguard the environment.[1]

In 2008, the agricultural professor Nathan Fiala* and the economics and environment scholar Fabio Grazi* published an article arguing that the ecological footprint addresses issues surrounding equality of resource distribution, but not environmental sustainability,[2] highlighting an important drawback in Wackernagel and Rees's model: it is not always consistent with the aim of maximizing social welfare.[3] Over the decade following its publication, however, other scholars have effectively improved the ecological footprint methodology.

Chad Monfreda,* for example, refined the model by differentiating between primary production (resource material generated by plants) and secondary production (resources generated by animal life), which led to more accurate calculations;[4] George Jenerette* (a scholar of the effects of land use and climate change)* attempted to combine the ecological footprint model with the values of environmental economics, seeking a middle ground between strong and weak approaches to sustainability; and the ecologist Michael Knaus worked out better estimates of the land area needed to balance human impacts on the environment.[5] In the end, the most important contribution of the book to the evolution of the field is its simple demonstration of the need to radically rethink economic models that depend on unlimited growth if we are going to move toward a sustainable future.

Schools of Thought

In the early to mid-1990s, an increasing number of scholars had begun to challenge traditional thinking about economics and global development, insisting that we cannot continue to build and consume in the same ways as before without endangering the survival of our species. This intellectual battlefield was well represented in the first chapter of *Our Ecological Footprint*.[6] These strong assertions were met by a number of opponents, most notably: those who think that free trade will solve our problems; believers in the power of technology to resolve ecological issues and the limited availability of energy sources; and deniers who simply want scientists to stop sending alarming messages to the public about catastrophes.

The strongest clash that occurred was between believers in the idea that permanent economic growth is sustainable if we find ways to absorb the environmental and social costs, and those who insist that we need a totally different economic system based on the acknowledgment of the Earth's limits and urgent measures to stop the depletion of natural capital.*

The first school of thought was dominant in the 1990s (and continues to prevail today). It is associated with the weak sustainability positions expressed by ecological and economic scholars such as Julian Simon,* Herman Kahn,* Warwick McKibbin,* and Jeffrey Sachs,* who all suggest technological and monetary fixes to overcome the ecological crisis.[7] The environmental economist David Pearce,* another affiliate of this school, maintains that the depletion of exhaustible resources happens by definition, so monetary valuation* of natural resources and ecological damage makes sense.[8] Wackernagel explicitly challenges Pearce's logic on the grounds that resource scarcity is seldom accurately reflected in financial markets.[9]

Wackernagel and Rees belong to the school of thought based on ecological economics (as opposed to environmental economics, which focuses on weak sustainability that accommodates constant economic growth), along with scholars such as John Cobb,* Herman Daly,* and Robert Costanza.* Daly and Cobb's work stresses that "once nature is overexploited, a loss of nature's services cannot be compensated by a gain in manufactured goods."[10]

Contrary to supporters of neoclassical economics,* ecological economists argue that human monetary capital fundamentally relies on natural capital, which must be protected for future generations. The innovative economic thinkers Kenneth E. Boulding* and Nicholas Georgescu-Roegen* are considered the founders of ecological economics. Boulding was a cocreator of General Systems Theory,* a principal component of evolutionary economics,* which draws parallels between economic and biological development. These theories mounted a powerful challenge to conventional economic theory. Georgescu-Roegen (who served as an advisor to Herman Daly) aimed to set economics on a foundation based on thermodynamics,* the branch of physics that shows the gradual decline of the energy in a system. *Our Ecological Footprint*, along with Wackernagel and Rees's other works, are clearly positioned in the field of ecological economics, and continue its critique of conventional economic principles.[11]

In Current Scholarship

The ideas put forward by Wackernagel and Rees have inspired a number of applications, among them calculations of regional and national ecological footprints;[12] case studies on the environmental impacts of commuting and housing; complex analyses of ecological deficits* and the ecological implications of international trade; educational projects; and calculations of individual ecological footprints, giving ordinary people a sense of their personal impact on the planet. The World Wildlife Fund*—a leading international conservation charity—uses ecological footprint calculations* in their advocacy efforts. The US-based think tank Redefining Progress uses this model to develop policies that balance economic well-being, environmental preservation, and social justice. At the governmental level, the European Union's Directorate General for the Environment has used a version of Wackernagel and Rees's methodology to calculate the environmental footprint of a variety of products.[13] The Scottish government has also adopted the ecological footprint model in their national performance framework, setting an ambitious target to cut carbon dioxide emissions by 80 percent by 2050.

The book's critique of permanent economic growth has not been as widely recognized as the ecological footprint accounting tool. For example, although the Scottish government has measured, and seeks to reduce, its ecological footprint, it also aims to "increase sustainable economic growth,"[14] which reflects a weak sustainability approach. This complex issue remains an arena for conflict, conversation, and innovation. In the academic debate, spatial welfare* economic analysis, which aims to maximise social welfare as the key component of sustainability strategies, offers an example of a new approach designed to counter the ecological footprint's limitations in this regard.[15]

Therefore, while direct opposition to Wackernagel and Rees's model remains, a greater number of scholars have used the text as a springboard and developed new ideas and techniques to build on or address the original concept's strengths and weaknesses.

NOTES

1   Robert Constanza et al., "Commentary Forum: The Ecological Footprint," *Ecological Economics* 32, no. 3 (2000).

2   N. Fiala, "Measuring Sustainability: Why the Ecological Footprint is Bad Economics and Bad Environmental Science," *Ecological Economics* 67, no. 4 (2008): 520.

3   F. Grazi et al., "Spatial Welfare Economics versus Ecological Footprint: Modeling Agglomeration, Externalities and Trade," *Environmental and Resource Economics* 38, no. 1 (2007): 151–2.

4   C. Monfreda et al., "Establishing National Natural Capital Accounts Based on Detailed Ecological Footprint and Biological Capacity Assessments," *Land Use Policy* 21, no. 3 (2004): 231–46.

5   M. Knaus et al., "Valuation of Ecological Impacts—A Regional Approach Using the Ecological Footprint Concept," *Environmental Impact Assessment Review* 26, no. 2 (2006).

6   Mathis Wackernagel and William Rees, *Our Ecological Footprint: Reducing Human Impact on the Earth* (Gabriola Island, British Columbia: New Society Publishers, 1996), 17–27.

7   J. L. Simon and H. Kahn, eds., *The Resourceful Earth: A Response to "Global 2000"* (New York: Wiley-Blackwell, 1984); Warwick J. McKibbin and Jeffrey D. Sachs, *Global Linkages: Macroeconomic Interdependence and Cooperation in the World Economy* (Washington, DC: Brookings Institution Press, 1991)

8   David W. Pearce and R. Kerry Turner, *Economics of Natural Resources and the Environment* (Baltimore, MD: Johns Hopkins University Press, 1990).

9   M. Wackernagel, "Ecological Footprint and Appropriated Carrying Capacity: A Tool for Planning Toward Sustainability" (PhD diss., University of British Columbia, 1994): 63–4.

10   H. E. Daly and J. Cobb, *For the Common Good: Redirecting the Economy Towards Community, the Environment and a Sustainable Future* (London: Green Print, 1990), 72.

11   For example, Rees published several earlier works in which he argued that traditional economic analysis cannot even understand the sustainability crisis: W. E. Rees, "The Ecology of Sustainable Development," *The Ecologist* 20, no. 1 (1990); W. E. Rees and M. Wackernagel, "Ecological Footprints and Appropriated Carrying Capacity: Measuring the Natural Capital Requirements of the Human Economy," in *Investing in Natural Capital: The Ecological Economics Approach to Sustainability*, eds. A. M. Jansson et al. (Washington, DC: Island Press, 1994), 362–90.

12   M. Wackernagel et al., "Ecological Footprint Time Series of Austria, The Philippines, and South Korea for 1961–1999: Comparing the Conventional Approach to an 'Actual Land Area' approach," *Land Use Policy* 21, no. 3 (2004).

13   "Product Environmental Footprint PEF," European Commission, accessed January 24, 2016, http://ec.europa.eu/environment/eussd/smgp/index.htm.

14   The Scottish Government, "A National Performance Framework," *Scottish Budget Spending Review 2007*, accessed January 24, 2016, http://www.gov. scot/Publications/2007/11/13092240/9.

15   F. Grazi et al., "Spatial Welfare Economics Versus Ecological Footprint: Modeling Agglomeration, Externalities and Trade," *Environmental and Resource Economics* 38, no. 1 (2007).

MODULE 11
IMPACT AND INFLUENCE TODAY

KEY POINTS

- *Our Ecological Footprint* is still widely debated in academic and public policy circles, and it remains influential in larger discussions about sustainability* and the reduction of human impact on the Earth.

- The book continues to challenge the assumption of conventional neoclassical economics,* showing that unlimited growth and sustainable development* are fundamentally incompatible.

- The debate between ecological economists* and environmental economists* persists, with the former group arguing for an acknowledgment of limits to growth and finite resources, and the latter advocating more environmentally sensitive economic growth and incorporating market values for ecological goods and services.

Position

Mathis Wackernagel and William Rees's *Our Ecological Footprint: Reducing Human Impact on the Earth* remains at the center of debates about both the principles and the methods involved in sustainability assessments, and continues to be the focus of sophisticated responses and critiques. Wackernagel and Rees define a sustainable ecological footprint* as a use of resources and a level of waste production inside the limits that an area can support. Use beyond this level is called overshoot* and is dangerous, threatening the availability of resources in the future. Another way to look at this is to consider sustainability as preserving natural capital* (things such as forests and fish stocks, which

> ❝ If each nation were to export only true surpluses— output in excess of local consumption whose export would not deplete self-producing natural capital stocks— then the net effect would be an ecologically steady-state and global stability. ❞
>
> Wackernagel and Rees, *Our Ecological Footprint: Reducing Human Impact on the Earth*

are renewable if they aren't over-exploited) and consuming only natural income (the harvests or yields from natural capital);* this distinction has been a central part of Wackernagel and Rees's later work.[1]

The challenge to the mainstream ideas of unlimited growth and the pursuit of a "green economy," as favored by environmental economics, is ongoing. New interpreters of the strong sustainability* approach, such as Jason Venetoulis* and John Talberth,* build on Wackernagel and Rees's model but take a fully eco-centric perspective* by including the resource needs of other organisms in their indicators.[2] Although *Our Ecological Footprint* is 20 years old, it keeps challenging environmental economists' ideas about natural capital substitution. According to the authors, "once nature has been over-exploited, no amount of manufactured goods will compensate for the loss of natural capital."[3]

Following the book's publication, Wackernagel and Rees produced a notably strong article explicitly addressing the controversies between the two economic models. They attacked mainstream claims, such as those based on the theory of utility which suggests that individuals always act to maximize their economic benefits, and they proposed using the precautionary principle.* This is an approach to respond to possible dangers to the health of humans, animals, plants, or the environment, to encourage action to stop ecological damage rather than waiting for certain debates to be resolved (by which time it may be too late).[4] The ecological footprint provides a valuable "warning light," exposing the potentially dangerous disparities between current resource use and long-term resource availability.

Interaction

Our Ecological Footprint remains central to the intellectual debate about environmental sustainability. Wackernagel and the Global Footprint Network's* interactions with national governments, cities, the financial sector, and local groups have expanded the conversation beyond academic circles and into the real world of public policy.[5] For instance, in 2012, a public debate between former Greenpeace* Executive Director Paul Gilding,* and the entrepreneur and philanthropist Peter Diamandis,* asked what we should do with a world full of people, goods, and waste.[6] Moreover, the ecological crises predicted by many scientists are now closer to the center of many governments' political agendas than they were 20 years ago. Steven Cork,* an ecologist for Australia's National Science Agency, concluded that in Australia "it is likely that consumption of resources will outstrip supply leading to a decline in the quality of life of Australians."[7] Increasingly, official discussions extend widely to include population, immigration, and lifestyle issues, and many use the ecological footprint concept as a central reference point.

In the ongoing scientific debate, responses to the book are mostly situated in the further development of ecological economics.* An ecological economics-based approach is supported by recent studies of the relationship between the ecological footprint and per capita Gross Domestic Product (GDP).* These studies suggest that "indefinite economic growth within a clean environment cannot be achieved simultaneously by the whole planet, since it can only work locally until there are countries whose environment is allowed to deteriorate."[8] *Our Ecological Footprint* argues that ecologically balanced trade could only be possible if "each nation were to export only true surpluses—output in excess of local consumption whose export would not deplete self-producing natural capital stocks—then the net effect would be an ecologically steady-state and global stability."[9] The recent adoption of the Paris Agreement* at the 21st Conference of the Parties (COP21) of the United Nations Framework Convention on Climate Change (UNFCCC) is testament to the fact that the reduction of human impacts on the planet, measurable via ecological and carbon footprints, is surging to the top list of governments' priorities.

The Continuing Debate

Our Ecological Footprint has been cited over 6,600 times by other scholars,[10] and its reach goes far beyond academic literature. It has been translated into eight languages, and remains a top seller in the field. The success of Wackernagel's Global Footprint Network (GFN) is exemplified by a $1,000,000 grant received from the Britain and China-based Skoll Foundation for Social Entrepreneurship in 2007. These funds have been used to institutionalize ecological footprint accounting in 10 countries, and the GFN works on sustainability projects with over 75 partners worldwide. The authors' stated intent "to make the case that we humans have no choice but to reduce our ecological footprint" has been achieved and the framework has resonated particularly at local and regional levels.[11]

There have also been several attempts to bridge the academic disputes between environmental economics and ecological economics. For example, the scholar of land use and climate change* George Jenerette* combined ecosystem* service valuation with ecological footprint analysis to create an integrated model.[12] Similarly, Michael Knaus, a scholar whose work draws on the fields of economics and ecology,* applied the ecological footprint concept jointly with indirect monetary valuation* to calculate offset areas that can compensate for land use impacts.[13] Despite these efforts, a large divide persists between the authors' school of thought (strong sustainability and ecological economics) and the weak sustainability* posture of neoclassical and environmental economists.

Venetoulis and Talberth changed the departure point of ecological footprint analysis from an anthropocentric*—or human–based— orientation to a broader ecologically based perspective.[14] They made useful transformations to the ecological footprint methodology by considering the entire Earth and other species in their accounts. These are important improvements that represent a positive evolution of the authors' original approach. Moreover, scholars such as Jenerette and Knaus have tried to merge ecological footprint and traditional economic valuation techniques, creating intellectual space for a dialogue between ecological and environmental economists.[15]

NOTES

1   M. Wackernagel et al., "National Footprint and Biocapacity Accounts 2005: The Underlying Calculation Method," *Global Footprint Network* 33 (2005).

2   J. Venetoulis and J. Talberth, "Refining the Ecological Footprint," *Environment Development and Sustainability* 10 (2008): 441.

3   Mathis Wackernagel and William Rees, *Our Ecological Footprint: Reducing Human Impact on the Earth* (Gabriola Island, British Columbia: New Society Publishers, 1996), 47.

4   M. Wackernagel and W. E. Rees, "Perceptual and Structural Barriers to Investing in Natural Capital: Economics from an Ecological Footprint Perspective," *Ecological Economics* 20, no. 1 (1997): 4.

5   "What We Do," Global Footprint Network, accessed January 24, 2016, http://www.footprintnetwork.org/en/index.php/GFN/.

6   Paul Gilding and Peter Diamandis, "Exclusive Q&A from the TED stage: Paul Gilding and Peter Diamandis debate," *TEDBlog*, March 14, 2012, accessed September 22, 2012, http://blog.ted.com/2012/03/14/exclusive-qa-from-the-ted-stage-paul-gilding-and-peter-diamandis-debate/.

7   S. Cork, "Ways Forward in the Population and Environment Debate," Parliament of Australia, Department of Parliamentary services, in *Pre-Election Policy Unit of the Parliamentary Library* (2010): I–II.

8   M. Bagliani et al., "A Consumption-Based Approach to Environmental Kuznets Curves Using the Ecological Footprint Indicator," *Ecological Economics* 65, no. 3 (2008): 15.

9   Wackernagel and Rees, *Our Ecological Footprint*, 130.

10   "Our Ecological Footprint," Google Scholar, accessed January 24, 2016, https://scholar.google.co.uk/ scholar?hl=en&q=Our+Ecological+Footprint+&btnG=&as_sdt=1%2C5&as_ sdtp=.

11   Wackernagel and Rees, *Our Ecological Footprint*, xi.

12   G. D. Jenerette et al., "Linking Ecological Footprints with Ecosystem Valuation in the Provisioning of Urban Freshwater," *Ecological Economics* 59, no. 1 (2006).

13   M. Knaus et al., "Valuation of Ecological Impacts—A Regional Approach Using the Ecological Footprint Concept," *Environmental Impact Assessment Review* 26, no. 2 (2006).

14   J. Venetoulis and J. Talberth, "Refining the Ecological Footprint," *Environment Development and Sustainability* 10 (2008): 441–69.

15   Jenerette et al., "Linking Ecological Footprints;" and Knaus et al., "Valuation of Ecological Impacts."

MODULE 12
WHERE NEXT?

KEY POINTS

- *Our Ecological Footprint* has inspired a number of potential applications and continues to arouse interest across a wide range of fields.

- Scholars continue to work with and modify the ecological footprint* approach, and the authors' associated nongovernmental organizations (NGOs) have sought to institutionalize the ecological footprint as a measurement for sustainability* at local and national levels.

- Wackernagel and Rees constructed a useful model that allows ordinary people to picture the significant environmental challenges we collectively face, and to help inform the changes necessary to ensure the ecological and social well-being of future generations.

Potential

Mathis Wackernagel and William Rees's *Our Ecological Footprint: Reducing Human Impact on the Earth* centers on human ecology,* environmental sciences, and economics. However, it has also been influential in psychology, behavioral studies, and other disciplines. For example, the computer scientist Jennifer Mankoff* proposed an approach that would integrate ecological footprint data into social networking sites to explore how they can help reduce personal energy consumption.[1] In architecture, Simon Guy* and Graham Farmer* explore alternative ecological design logics and how they could be translated in specific technological strategies.[2] Similarly, in the field of psychology, Kirk Brown* and Tim Kasser* have argued that personal values and mindfulness (roughly, conscious awareness of our actions)

> **❝** As things stand, the pace of stock depletion and accelerating global change suggests that remaining natural capital stocks are already inadequate to ensure long-term ecological sustainability. In these circumstances, we believe that 'strong sustainability' is a necessary condition for ecologically sustainable development. **❞**
>
> Wackernagel and Rees, *Our Ecological Footprint: Reducing Human Impact on the Earth*

can be cultivated in young people to promote both personal well-being and ecologically responsible behaviors (measured by the ecological footprint).[3]

Although these are only a few examples of Wackernagel and Rees's influence outside their own field, they are especially significant given the growing prevalence of social media, regulations and investment into sustainable construction and design, and the importance of young people to determine the future direction of ecological action.

Wackernagel and Rees are emphatic that changing our consumption and waste-producing behavior is essential to confronting effectively the unfolding global environmental crisis, advocating a "shift from 'managing resources' to 'managing ourselves'."[4] Given the wide range of potential applications and the ongoing interest across numerous fields, the text clearly retains the potential to influence a diverse group of scholars and public discourse about environmental choices.

Future Directions

Our Ecological Footprint's emphasis on the power of human behavior to cause serious environmental collapse or to create positive changes toward more sustainable societies will remain central to the debates and challenges ahead. The wide-ranging scope of initiatives launched in the two decades following the book's publication demonstrates its ongoing power to influence new generations of students, scientists, and decision makers. Some cities and communities have made substantial changes in the direction of sustainability as a result of their work with the Global Footprint Network* and One Earth* (NGOs run and founded by Wackernagel and Rees respectively). However, overpopulation and over-consumption continue to threaten the environment.

It is likely that the book's core ideas will continue to develop thanks to a growing wealth of supportive scholarly publications, a proliferation of NGOs, national and international environmental reports, and increased media attention.[5] In a video featured on his website, William Rees observes that humans are by nature complacent and accustomed to their circumstances, and that only the threat of "climate catastrophe" might make us react and use our intellectual resources to address the environmental crises. However, he warns that tribalism may prevail during "crunch time," and stresses that cross-cultural communication is needed to revise our cultural myths in order to foster productive international cooperation for the global greater good.[6]

Correspondingly, Wackernagel has recently promoted "good bookkeeping of our resources," and he highlights the importance of "slow things," objects and ideas with long life cycles, such as housing and infrastructure, because they maintain their eco-friendly features for decades.[7] Studies on human behavior in times of crisis and in relation to urban planning have been among the recent applications helping to push the evolution of Wackernagel and Rees's model. This intellectual dynamism will likely continue because the sustainability problem remains both urgent and stubbornly unresolved. The authors themselves will carry this work forward through their NGOs and fresh collaborations with like-minded colleagues. Both have significant allies within the academic community—particularly in the field of ecological economics*—as well as public and institutional support—mainly from sympathetic environmentalists, and increasingly in local communities and international organizations using the ecological footprint to improve sustainability.

Summary

Our Ecological Footprint deserves special attention, especially from students, for its creation of a powerful and useful representation of humankind's environmental impact, accessible to everyone. In particular, the authors' forceful arguments and evocative illustrations invite readers to consider their behavior and collective decision-making in terms of sustainability. Wackernagel and Rees's book spurred important debates in ecological economics and beyond, and serves as a useful reminder of our fundamental dependence on the Earth's limited resources.

Importantly, the work established the tool of the "ecological footprint," which scholars continue to refine and which several national governments have adopted to assess their country's sustainability.

While *Our Ecological Footprint* was a major milestone in the careers of its authors, they each have made a significant impact beyond this particular work. William Rees developed concepts such as "regional capsule"* and "personal planetoid,"* which, after 20 years of research and teaching, were consolidated into the ecological footprint model. He has also made important contributions to human ecology and global social dynamics.[8] Rees recently developed a thesis advocating drastic reductions in global consumption and waste production through a proposal that would "tax the bads (depletion and pollution) not the goods (labor and capital)."[9]

Under William Rees's academic supervision, Mathis Wackernagel took up Rees's ideas and developed ecological footprint analysis as part of his doctoral dissertation, and eventually founded the Global Footprint Network, which advocates and applies the concept widely. He continued his research through collaborations with colleagues across the world, improved the ecological footprint methodology, and promoted its adoption by national and local governments.[10]

Ultimately, Wackernagel and Rees constructed a model that helps us visualize the significant environmental challenges we collectively face, as well as a tool that promotes the changes we urgently need to make if we are to meet the needs both of future generations and of the planet itself.

NOTES

1   J. Mankoff et al., "Leveraging Social Networks to Motivate Individuals to Reduce Their Ecological Footprints," *Human-Computer Interaction Institute,* Paper 47 (2007).

2   S. Guy and G. Farmer, "Reinterpreting Sustainable Architecture: The Place of Technology," *Journal of Architectural Education* 54, no. 3 (2001).

3   K. W. Brown and T. Kasser, "Are Psychological and Ecological Well-being Compatible? The Role of Values, Mindfulness, and Lifestyle," *Social Indicators Research* 74, no. 2 (2005): 349.

4   Mathis Wackernagel and William Rees, *Our Ecological Footprint: Reducing Human Impact on the Earth* (Gabriola Island, British Columbia: New Society Publishers, 1996), 4.

5   Paul Gilding and Peter Diamandis, "Exclusive Q&A from the TED stage: Paul Gilding and Peter Diamandis debate," *TEDBlog*, March 14, 2012, March 14, 2012, accessed September 22, 2012, http://blog.ted.com/2012/03/14/exclusive-qa-from-the-ted-stage-paul-gilding-and-peter-diamandis-debate/.

6   William Rees, "Why is Humanity in Denial?," Ecofootnotes, July 7, 2012, accessed January 24, 2016, www.williamrees.org/why-is-humanity-in-denial/.

7   Beppe Grillo, "The interviews of the www.beppegrillo.it blog: Mathis Wackernagel," accessed January 24, 2016, http://www.youtube.com/watch?v=_-QkH1sYwRuU&playnext=1&list=PLD897815F2DB1A247&feature=results_video.

8 "William Rees—Short Biography," University of British Columbia, School of Community and Regional Planning, accessed September 18, 2012, http://www.scarp.ubc.ca/people/william-rees.

9 W. E. Rees, "The Way Forward: Survival 2100," *Solutions* 3, no. 3 (2012).

10 "Partner Network," Global Footprint Network, accessed January 24, 2016, http://www.footprintnetwork.org/en/index.php/GFN/page/partner_network/.

GLOSSARY

GLOSSARY OF TERMS

Anthropocentric: a view of humankind as the center of existence, as opposed to God or animals.

Appropriated carrying capacity: a term that has become synonymous with the "ecological footprint," but refers more specifically to the importation of ecological capacity from distant places.

Biocapacity: the capacity of ecosystems to produce an ongoing supply of renewable resources and to absorb waste; if an ecosystem's biocapacity is exceeded, it becomes unsustainable.

Biodiversity: the variety and variability of living organisms that conservationists aim to protect.

Biophysics: the science that applies physics in order to understand biological phenomena and processes.

Biosphere: the part of the Earth and its atmosphere where organisms, such as plants and animals, live.

Brundtland Commission: also known as the United Nations World Commission on Environment and Development, this was a forum convened in 1983 to study strategies for achieving sustainable development. It was dissolved in December 1987 after it released a major report of its findings: *Our Common Future.*

Capitalism: an economic system in which privately owned goods and services are exchanged for profit.

Carbon footprint: part of an individual, business, or geographic area's total ecological footprint, this represents the total amount of carbon dioxide and other carbon compounds emitted to directly or indirectly support human activities—notably through the burning of fossil fuels. The phrase is frequently used in the wider climate-change debate.

Carrying capacity: defined by Wackernagel and Rees as "the maximum population size of a given species that an area can support without reducing its ability to support the same species in the future. In the human context, the social scientist William Catton defines it as the maximum 'load' that can safely and persistently be imposed on the environment by people."

Climate change: change in weather patterns and average global temperatures over a period of time.

Club of Rome is an international think tank founded in 1968 in Rome. It is most famous for its publication *Limits to Growth* (1972), which was based on a computer model (World3) simulating the ways in which growth of economy and population determined the consumption of natural resources. The authors of *Limits to Growth* are Donella H. Meadows, Dennis L. Meadows, Jørgen Randers, and William W. Behrens III.

Community and regional planning: a professional field concerned with helping communities develop the local economy and infrastructure.

Eco-centric perspective: a point of view that places the ecosphere, rather than the biosphere, at the center of attention thus trying to counterbalance anthropocentrism.

Ecological accounting: the use of accounting and finance models to estimate the environmental costs of economic decisions.

Ecological deficit: a situation whereby the ecological footprint of a population exceeds the biocapacity of the area in which that population lives.

Ecological economics: a field that bridges the disciplines of ecology and economics, routinely drawing on those of psychology, anthropology, archaeology, and history.

Ecological footprint: defined by Wackernagel's Global Footprint Network as "a measure of how much biologically productive land and water an individual, population, or activity requires to produce all the resources it consumes and to absorb the waste it generates."

Ecology: a field of study related to biology that focuses on the relationship between organisms and their environment.

Ecosphere: the planetary ecosystem, consisting of all living organisms and their environment.

Ecosystem: the complex set of interrelationships between and among living organisms and their physical environment.

Energy land: the sum of all areas used to provide crops and wood for fuel and hydroelectric power energy.

Entropy: a measure of the energy that is not available to do work in a closed thermodynamic system, as per the second law of thermodynamics, which states that an isolated system will degenerate into a more disordered state.

Environmental economics: a branch of economics that analyzes financial impacts from environmental policies, such as regulatory compliance costs.

Evolutionary economics: a subfield of economics that draws its inspiration from evolutionary biology.

Footprint calculators: algorithms allowing people to calculate their per-capita impact on the planet (for example, their ecological footprint).

Fossil fuel: living matter accumulated over geological times, such as hydrocarbon deposits—oil, coal, or natural gas—extracted and used for fuel.

General Systems Theory: refers to a level of theoretical model building that is situated between the generalized construction of pure mathematics and the theories of the specific fields.

Geographical Information System (GIS): a data management system to store and manipulate geographical information.

Global Footprint Network: a nongovernmental organization established by Mathis Wackernagel in 2003 that focuses on international sustainability.

Globalization: a process of international integration; such integration takes many forms, including economic, political, and cultural.

Greenpeace: an independent campaigning organization founded in 1971 that exposes global environmental problems and promotes solutions, particularly by means of peaceful protest and creative communication.

Gross Domestic Product (GDP): the market value of all the goods and services produced by labor and property located in a country or region.

Human carrying capacity: the maximum number of people or organisms (plants or animals) that can be sustained in a given area without destroying the local environment.

Human ecology: the study of the relationships between humans and nature.

Input–output (IO/I–O) analysis: a mathematical tool used in economics to analyze how goods and services flow between sectors. I–O tables are used under the assumption of full consumption of an industry's products by consumers or other industries and product traceability.

Internalization of environmental costs: a way of accounting for the cost of environmental impacts (for instance the effects of air pollution on people's health) so that environmentally unfriendly activities, such as fossil-fuel use, become more expensive.

International trade theory: a theory that explains the pattern of international trade and distribution of its positive outcomes, and highlights the benefits of liberal trade.

Localism: devotion to and promotion of the interests of a particular locality.

Monetary valuation: a process that associates currency units with services provided by the natural environment, or damages done to it.

Natural capital: the natural assets providing valuable goods and services to people. Its sustainable harvest is "natural income" (that is, one that is replenished by the Earth's regenerative capacity).

Neoclassical economics: a school of economics that constructs its understanding of markets and resource allocation based on three core assumptions: people are rational and have discernible preferences based on value; people maximize utility while firms maximize profits; people act with full and relevant information.

Neoliberalism: a political and economic doctrine in favor of privatization, free trade, and reduced government interference in business, along with low public expenditure on social services.

One Earth: a Canadian research and advocacy group that seeks to foster long-term sustainability.

Overshoot: the use of nature's resources beyond their limits.

Personal planetoid: term, defined by William Rees, as "the per capita ecological footprint."

Paris Agreement: a major international agreement adopted in Paris by consensus between 197 countries on December 12, 2015; it sets out measures to reduce greenhouse gas emissions after 2020, fund and implement adaption to the impacts of climate change, and address loss and damage demands associated with adverse effects of climate change, such as extreme weather events. It will enter into force if and when 55 countries responsible for the production of at least 55% of the world's greenhouse emissions ratify, accept, approve, or accede to the agreement.

Photosynthesis: the process through which plants convert sunlight into energy.

Population ecology: a subfield of ecology concerned with the study of the distribution, dynamics, and structure of animal and plant populations.

Precautionary principle: an approach defined in Principle 15 of the 1992 Rio Earth Summit that "enables rapid response in the face of a possible danger to human, animal, or plant health, or to protect the environment. In particular, where scientific data do not permit a complete evaluation of the risk, recourse to this principle may be used to stop [a practice that might] be hazardous."

Regional capsule: a precursor to the "ecological footprint," this was an idea developed by William Rees in the 1970s to demonstrate that cities are sustained by lands that far exceed urban boundaries. The capsule is a metaphor that attempts to visualize the entire land area needed to sustain urban life.

Second law of thermodynamics: commonly referred to as "entropy," this concept holds that the total energy of a system tends to decrease if no energy enters or leaves the system.

Spatial welfare analysis: an approach aimed to maximise social welfare as the key component of sustainability strategies.

Steady-state economics: a stable condition with no change over time, or in which changes in different directions constantly balance one another.

Strong sustainability: an approach to sustainability that finds unlimited economic growth and ecological conservation as fundamentally incompatible, and demands a shift away from conventional neoclassical economic models.

Sustainability: strategies to use natural resources in a way that allows them to continue to renew themselves.

Sustainability planning: devising strategies to help develop infrastructure and economic models that focus on ensuring resources will continue to be available to future generations.

Sustainable development: development that satisfies the needs of the present generation while preserving those of future generations.

Thermodynamics: physics of the relationships and conversions between various forms of energy, including heat and entropy.

United Nations: an international organization of countries established in 1945 with the aim of promoting peace, security, and cooperation globally.

Weak sustainability: an approach to sustainability that holds that society and environment are the dimensions between which to find compromise in order to maximize our socioeconomic benefits while minimizing the impacts on the environment.

World Wildlife Fund: a nongovernmental organization that promotes biodiversity conservation and limiting the environmental impact that humans have on the planet.

Zero-sum: a concept used in international relations to describe a situation where a victory for one side of a conflict is viewed as a loss for the other.

PEOPLE MENTIONED IN THE TEXT

Jeroen van den Bergh (b. 1965) is a Dutch environmental economist. He is best known for his work on the economics of sustainability.

Kenneth Ewart Boulding (1910–93) was an economist, educator, poet, systems scientist, interdisciplinary philosopher, peace activist, and devoted member of the nonconformist Christian group known as the Quakers.

Kirk Brown is an assistant professor of psychology at Virginia Commonwealth University.

Erwin Bulte is a Dutch professor of development economics at Wageningen University.

Susan Burns is chief executive officer and founder of Global Footprint Network, along with her husband, Mathis Wackernagel.

Rachel Carson (1907–64) was an ecologist and writer who courageously spoke out against pollution of ecosystems by fertilizers and pesticides in the US. In 1962, she wrote the milestone book *Silent Spring*; the following year she called for new policies to protect human health and the environment in Congress. At the time she was attacked as an alarmist by the chemical industry and some in government, but has since become an iconic figure for those campaigning for environmental conservation.

William Catton Jr. (1926–2015) was a US sociologist best known for his research in environmental sociology on topics such as carrying capacity.

Nicky Chambers is the co-founder of Best Foot Forward, a nongovernmental organization that consults on sustainability, particularly on the question of carbon and ecological footprinting.

John B. Cobb Jr. (b. 1925) is an American United Methodist theologian. He collaborated with Herman Daly in writing *For the Common Good: Redirecting the Economy Towards Community, the Environment and a Sustainable Future* (1989).

Steven Cork is an adjunct professor at the Australian National University and a consultant on issues related to ecology.

Robert Costanza (b. 1950) is an ecological economist and currently chair in public policy at the Crawford School of Public Policy, Australian National University. Costanza's research deals with research, policy, and management issues regarding humans–nature interactions at small to large time and space scales.

Herman E. Daly (b. 1938) is an emeritus professor at the University of Maryland, School of Public Policy, and former senior economist in the Environment Department of the World Bank (1988–94). His main research interests are sustainable economic development, resources, and population.

Peter Diamandis (b. 1961) is a US entrepreneur who is best known for founding the X Prize, awarded to an individual or organization that develops technology that benefits all mankind.

Paul R. Ehrlich (b. 1932) is a US scientist who has been pioneering in raising awareness over the crucial issues of overpopulation and sustainability.

Salah El Serafy is a senior economist and adviser to the World Bank who often deals with ecological issues.

Graham Farmer is a professor of architecture at Newcastle University, whose research focuses on sustainable architecture.

J. Bruce Falls is a Canadian naturalist and zoologist, and founding member of the Nature Conservancy of Canada.

Nathan Fiala is an assistant professor of agriculture at the University of Connecticut.

Mario Giampietro (b. 1953) is an Italian-born research professor at the Universitat Autònoma de Barcelona, whose research focuses on sustainability.

Nicholas Georgescu-Roegen (1906–94) was a Romanian American mathematician, statistician, and economist. His most renowned work is *Entropy Law and the Economic Process* (1971), which proposed that the Second Law of Thermodynamics on entropy governs the economy.

Peter Gilding is an Australian-born environmental activist, best known for his role as the executive director of Greenpeace.

Fabio Grazi is an economist and researcher at the Agence Française de Développement (AFD), whose research focuses on the relationship between economic development and the environment.

Simon Guy is a professor of architecture and planning at Lancaster University, where he is also dean of arts and social sciences. His research focuses on sustainable urbanism.

George Jenerette is an assistant professor in the department of botany and plant sciences at Arizona State University. His research focuses on how land use and climate changes can affect ecosystems.

Tim Kasser (b. 1966) is a professor of psychology at Knox College with wide-ranging research interests. He is known for an article that he coauthored with Tom Crompton called *Human Identity: A Missing Link in Environmental Campaigning*, which dealt with the psychology of climate-change denial.

Herman Kahn (1922–83) was a research analyst and military strategist, who went on to found the Hudson Institute.

Justin Kitzes is a theoretical and applied ecologist at the University of California, Berkeley, whose research focuses on how to predict the impact of changes on biodiversity.

G. Cornelis van Kooten is a professor of economics at the University of Victoria, Canada, whose research focuses on natural resource economics.

Roger Levett is a sustainability consultant and cofounder of the consultancy, Levett-Therivel.

Jennifer Mankoff is an associate professor of computer sciences at Carnegie Mellon University, whose research often deals with environmental sustainability.

Warwick McKibbin (b. 1957) is an Australian-born professor of economics at the Australian National University, whose research focuses on global economic modeling.

Ian Moffatt is a professor of sustainable development at the University of Stirling, UK, whose research focuses on modeling and measuring sustainable development.

Chad Monfreda is a postdoctoral fellow at the Gerald R. Ford School of Public Policy at the University of Michigan, whose research focuses on the carbon markets in California and Mexico.

David Pearce (1942–2005) was a pioneer in environmental economics and professor at University College London (UCL) from 1983; he developed pricing methods for factoring environmental damages into the economy.

David Pimentel is a professor emeritus at Cornell University, whose research focuses on biodiversity, ecology, and environmental sciences.

François Quesnay (1694–1774) was the leading figure of the Physiocratic School, generally considered to be the first school of economic thinking.

David Rapport is a Canadian American professor of resource economics at Royal Roads University in Victoria, Canada.

Jeffrey Sachs (b. 1954) is an American economist, who serves as the director of the Earth Institute at Columbia University. He is best known for his best-selling books *The End of Poverty* (2005), *Common Wealth* (2008), and *The Price of Civilization* (2011).

Craig Simmons is the cofounder of Best Foot Forward, a nongovernmental organization that consults on sustainability, particularly on the question of carbon and ecological footprinting.

Judith Silverstein is an associate of Mathis Wackernagel; the two cowrote the 2002 article "Big things first: focusing on the scale imperative with the ecological footprint."

Julian L. Simon (1932–98) was an economics professor at the University of Illinois and a business administration professor at the University of Maryland. He wrote *The Ultimate Resource* (1981) and *The Ultimate Resource II* (1996) to put forward the view that humans will indefinitely be able to harness resources and energy from nature overcoming limits and obstacles with ingenuity and innovation.

John Talberth is an environmental economist and the founder of the Center for Sustainable Economy, whose research focuses on the economics of sustainable development.

Jason Venetoulis is an environmental economist, whose research focuses on the ecological footprint and biocapacity.

Heroen Verbruggen is a marine biologist at the University of Melbourne, Australia, whose research focuses on the evolutionary diversification of marine algae.

Peter Morrison Vitousek (b. 1949) is an American professor of ecology at Stanford University, whose research focuses on the nitrogen cycle and on human appropriation of the products of photosynthesis.

WORKS CITED

WORKS CITED

Ayres, R. U. "Commentary on the Utility of the Ecological Footprint Concept." *Ecological Economics* 32, no. 3 (2000): 347–9.

Bagliani, M., G. Bravo, and S. Dalmazzone. "A Consumption-Based Approach to Environmental Kuznets Curves Using the Ecological Footprint Indicator." *Ecological Economics* 65, no. 3 (2008): 650–61.

Bergh, J. C. J. M. van den, and H. Verbruggen. "Spatial Sustainability, Trade and Indicators: An Evaluation of the 'Ecological Footprint'." *Ecological Economics* 29 (1999): 61–72.

Bond, S. *Ecological Footprints: A Guide for Local Authorities*. Godalming: World Wildlife Fund UK, 2002.

Brown, K. W., and T. Kasser. "Are Psychological and Ecological Well-Being Compatible? The Role of Values, Mindfulness, and Lifestyle." *Social Indicators Research* 74, no. 2 (2005): 349–68.

Brown, L. R. *Facing Food Insecurity*. In *State of the World 1994: A Worldwatch Institute Report on Progress Toward a Sustainable Society*, edited by L. R. Brown et al., 177–97. New York: W. W. Norton & Company, 1994.

Carson, R. *Silent Spring*. New York: Mariner Books, 2002.

Catton, William R. Jr. *Overshoot: The Ecological Basis of Revolutionary Change*. Urbana: University of Illinois Press, 1980.

Chambers, N., C. Simmons, and M. Wackernagel. *Sharing Nature's Interest: Ecological Footprints as an Indicator of Sustainability*. London: Earthscan, 2000.

Cork, S. "Ways Forward in the Population and Environment Debate." Parliament of Australia, Department of Parliamentary Services. In *Pre-Election Policy Unit of the Parliamentary Library*, 2010.

Costanza, R., et al. "Commentary Forum: The ecological footprint." *Ecological Economics* 32, no. 3 (2000): 337–94.

Costanza, R. "The Dynamics of the Ecological Footprint Concept." *Ecological Economics* 32, no. 3 (2000): 341–5.

Daly, Herman. E. *Steady-State Economics*. Second Edition with New Essays. Washington, DC: Island Press, 1991.

Daly, Herman E., and John B. Cobb Jr. *For the Common Good: Redirecting the Economy Towards Community, the Environment, and a Sustainable Future*. London: Green Print, 1990.

Ehrlich, Paul R. "Human Carrying Capacity, Extinctions, and Nature Reserves." *BioScience* 32, no. 5 (1982): 331–3.

El Serafy, Salah. "The Proper Calculation of Income from Depletable Natural Resources." In *Environmental Accounting for Sustainable Development*, edited by Yusuf J. Ahmad, Ernst Lutz and Salah El Serafy, 10–18. Washington, DC: The World Bank, 1989.

Fiala, N. "Measuring Sustainability: Why the Ecological Footprint is Bad Economics and Bad Environmental Science." *Ecological Economics* 67, no. 4 (2008): 519–25.

Giampietro M., and D. Pimentel. "Energy Efficiency: Assessing the Interaction Between Humans and Their Environment." *Ecological Economics* 4, no. 2 (1991): 117–44.

Global Footprint Network. "Europe 2007 Gross Domestic Product and Ecological Footprint." Accessed January 24, 2016. http://www.footprintnetwork.org/images/uploads/europe_2007_gdp_and_ef.pdf.

Grazi, F., J. C. J. M. van den Bergh, and P. Rietveld. "Spatial Welfare Economics Versus Ecological Footprint: Modeling Agglomeration, Externalities and Trade." *Environmental and Resource Economics* 38, no. 1 (2007): 135–53.

Guy, S., and G. Farmer. "Reinterpreting Sustainable Architecture: The Place of Technology." *Journal of Architectural Education* 54, no. 3 (2001): 140–8.

Hardin, G. "The Tragedy of the Commons." *Science* 162, no. 3 (1968): 1243–8.

Hunter, C., and J. Shaw. "The Ecological Footprint as a Key Indicator of Sustainable Tourism." *Tourism Management* 28, no. 1 (2007): 46–57.

Illge, L., and R. Schwarze. "A Matter of Opinion: How Ecological and Neoclassical Environmental Economists Think About Sustainability and Economics." Berlin: German Institute for Economic Research, 2006.

International Fund for Agricultural Development. "Combating Environmental Degradation." Accessed January 24, 2016. http://www.ifad.org/events/past/hunger/envir.html.

Jenerette, G. D., W. A. Marussich, and J. P. Newell. "Linking Ecological Footprints with Ecosystem Valuation in the Provisioning of Urban Freshwater." *Ecological Economics* 59, no. 1 (2006): 38–47.

Kitzes, J., et al. "A Research Agenda for Improving National Ecological Footprint Accounts." *Ecological Economics* 68, no. 7 (2009): 1991–2007.

Knaus, M., D. Löhr, and B. O'Regan. "Valuation of Ecological Impacts—a Regional Approach Using the Ecological Footprint Concept." *Environmental Impact Assessment Review* 26, no. 2 (2006): 156–69.

Kooten, G. C. van, and E. H. Bulte. "The Ecological Footprint: Useful Science or Politics?" *Ecological Economics* 32, no. 3 (2000): 385–9.

Lang, C. "Vested Interests: Industrial Logging and Carbon in Tropical Forests." REDDmonitor, June 26, 2009. Accessed January 24, 2016. http://www.redd-monitor.org/2009/06/26/vested-interests-industrial-logging-and-carbon-in-tropical-forests/.

Lenzen, M., and S. A. Murray. "A Modified Ecological Footprint Method and its Application to Australia." *Ecological Economics* 37, no. 2 (2001): 229–55.

Levett, R. "Footprinting: A Great Step Forward, but Tread Carefully—a Response to Mathis Wackernagel." *Local Environment: The International Journal of Justice and Sustainability* 3, no. 1 (1998): 67–74.

Mankoff, J., D. Matthews, S. R. Fussell, and M. Johnson. "Leveraging Social Networks to Motivate Individuals to Reduce Their Ecological Footprints." 40th Annual Hawaii International Conference on System Sciences, 2007.

McKibbin, Warwick J., and Jeffrey D. Sachs. *Global Linkages: Macroeconomic Interdependence and Cooperation in the World Economy*. Washington, DC: Brookings Institution Press, 1991.

Meadows, D. H., Dennis L. Meadows, Jørgen Randers, and William W. Behrens. *The Limits to Growth: A Report for the Club of Rome's Project on the Predicament of Mankind*. 2nd edition. New York: Universe Books, 1974.

Milbrath, L. W. *Envisioning a Sustainable Society: Learning Our Way Out*. Albany, NY: State University of New York Press, 1989.

Moffatt, I. "Ecological Footprints and Sustainable Development." *Ecological Economics* 32, no. 3 (2000): 359–62.

Monfreda, C., M. Wackernagel, and D. Deumling. "Establishing National Natural Capital Accounts Based on Detailed Ecological Footprint and Biological Capacity Assessments." *Land Use Policy* 21, no. 3 (2004): 231–46.

The Natural Capital Declaration. "The Declaration." Accessed January 24, 2016. http://www.naturalcapitaldeclaration.org/the-declaration/.

Opschoor, H. "The Ecological Footprint: Measuring Rod or Metaphor." *Ecological Economics* 32, no. 3 (2000): 363–5.

Pearce, David W., and R. Kerry Turner. *Economics of Natural Resources and the Environment*. Baltimore, MD: Johns Hopkins University Press, 1989.

Rapport, D. J. "Ecological Footprints and Ecosystem Health: Complementary Approaches to a Sustainable Future." *Ecological Economics* 32, no. 3 (2000): 367–70.

Rees, W. E. "Achieving Sustainability: Reform or Transformation?" *Journal of Planning Literature* 9 (1995): 343.

"Eco-Footprint Analysis: Merits and Brickbats." *Ecological Economics* 32, no. 3 (2000): 371–4.

"The Ecology of Sustainable Development." *The Ecologist* 20, no. 1 (1990): 18–23.

"How to Convince People to Face Reality." Citizen Action Monitor, September 22, 2011. Accessed January 24, 2016. http://citizenactionmonitor.wordpress.com/2011/09/22/william-rees-we-need-a-mass-social-movement-but-time-is-not-on-our-side/.

"Is Humanity Fatally Successful?" *Journal of Business Administration and Policy Analysis* (2004): 67–100.

"The Way Forward: Survival 2100." *Solutions* 3, no. 3 (2012).

Sachs, W. "The Gospel of Global Efficiency." *India International Centre Quarterly* 15, no. 3 (1988): 21–8.

Safire, W. "On Language: Footprint." *New York Times Magazine*, February 17, 2008.

The Scottish Government. "A National Performance Framework." *Scottish Budget Spending Review 2007*. Accessed January 24, 2016. http://www.gov.scot/Publications/2007/11/13092240/9.

Simon, J. L., and Kahn H., eds. *The Resourceful Earth: A Response to "Global 2000."* New York: Wiley Blackwell, 1984.

Sturm, A., M. Wackernagel, and K. Muller. *The Winners and Losers in Global Competition: Why Eco-efficiency Reinforces Competitiveness—A Study of 44 Nations.* West Lafayette, IN: Purdue University Press, 2003.

United Nations Conference on Environment and Development. "The Rio Declaration on Environment and Development." United Nations Environment Programme. Accessed January 24, 2016. http://www.unep.org/documents.multilingual/default.asp?documentid=78&articleid=1163.

Venetoulis, J., and J. Talberth. "Refining the Ecological Footprint." *Environment, Development and Sustainability* 10, no. 4 (2008): 441–69.

Vitousek, P. M., P. R. Ehrlich, A. H. Ehrlich, and P. A. Matson. "Human Appropriation of the Products of Photosynthesis." *BioScience* 36, no. 6 (1986): 368–73.

Wackernagel, M. "Ecological Footprint and Appropriated Carrying Capacity: A Tool for Planning Toward Sustainability." PhD diss., University of British Columbia, 1994.

Wackernagel, M., C. Monfreda, K. H. Erb, H. Haberl, and N. B. Schulz. "Ecological Footprint Time Series of Austria, the Philippines, and South Korea for 1961–1999: Comparing the Conventional Approach to an 'Actual Land Area' Approach." *Land Use Policy* 21, no. 3 (2004): 261–9.

Wackernagel, M., C. Monfreda, D. Moran, P. Wermer, S. Goldfinger, D. Deumling, and M. Murray. "National Footprint and Biocapacity Accounts 2005: The Underlying Calculation Method." *Global Footprint Network* 33 (2005).

Wackernagel M., L. Onisto, P. Bello, A. Callejas Linares, I. S. López Falfán, J. Méndez García, A. I. Suárez Guerrero, and M. G. Suárez Guerrero. "National Natural Capital Accounting with the Ecological Footprint Concept." *Ecological Economics* 29, no. 3 (1999): 375–90.

Wackernagel, Mathis, and William Rees. *Ecological Footprints and Appropriated Carrying Capacity: Measuring the Natural Capital Requirements of the Human Economy* (Vancouver: University of British Columbia, School of Community and Regional Planning, 1992).

Our Ecological Footprint: Reducing Human Impact on the Earth. Gabriola Island, British Columbia: New Society Publishers, 1996.

"Perceptual and Structural Barriers to Investing in Natural Capital: Economics from an Ecological Footprint Perspective." *Ecological Economics* 20, no. 1 (1997): 3–24.

Wackernagel, M., and J. Silverstein. "Big Things First: Focusing on the Scale Imperative with the Ecological Footprint." *Ecological Economics* 32, no. 3 (2000): 391–4.

Watson, R. T., Zinyowera, M. C. and Moss, R. H. *Climate Change 1995: Impacts, Adaptations and Mitigation of Climate Change: Scientific–Technical Analyses.* Cambridge: Cambridge University Press, 1996.

Weber, P. "Chapter 3: Safeguarding Oceans." In *State of the World 1994: A Worldwatch Institute Report on Progress Toward a Sustainable Society*, edited by L. R. Brown et al., 41–60. New York: W. W. Norton & Company, 1994.

Wiedmann, T., and J. Barrett. "A Review of the Ecological Footprint Indicator— Perceptions and Methods." *Sustainability* 2, no. 6 (2010): 1645–93.

World Wildlife Fund. "Living Planet Report." Gland, Switzerland, 2006. Accessed January 11, 2016. www.panda.org/livingplanet.

THE MACAT LIBRARY
BY DISCIPLINE

AFRICANA STUDIES

Chinua Achebe's *An Image of Africa: Racism in Conrad's Heart of Darkness*
W. E. B. Du Bois's *The Souls of Black Folk*
Zora Neale Huston's *Characteristics of Negro Expression*
Martin Luther King Jr's *Why We Can't Wait*
Toni Morrison's *Playing in the Dark: Whiteness in the American Literary Imagination*

ANTHROPOLOGY

Arjun Appadurai's *Modernity at Large: Cultural Dimensions of Globalisation*
Philippe Ariès's *Centuries of Childhood*
Franz Boas's *Race, Language and Culture*
Kim Chan & Renée Mauborgne's *Blue Ocean Strategy*
Jared Diamond's *Guns, Germs & Steel: the Fate of Human Societies*
Jared Diamond's *Collapse: How Societies Choose to Fail or Survive*
E. E. Evans-Pritchard's *Witchcraft, Oracles and Magic Among the Azande*
James Ferguson's *The Anti-Politics Machine*
Clifford Geertz's *The Interpretation of Cultures*
David Graeber's *Debt: the First 5000 Years*
Karen Ho's *Liquidated: An Ethnography of Wall Street*
Geert Hofstede's *Culture's Consequences: Comparing Values, Behaviors, Institutes and Organizations across Nations*
Claude Lévi-Strauss's *Structural Anthropology*
Jay Macleod's *Ain't No Makin' It: Aspirations and Attainment in a Low-Income Neighborhood*
Saba Mahmood's *The Politics of Piety: The Islamic Revival and the Feminist Subject*
Marcel Mauss's *The Gift*

BUSINESS

Jean Lave & Etienne Wenger's *Situated Learning*
Theodore Levitt's *Marketing Myopia*
Burton G. Malkiel's *A Random Walk Down Wall Street*
Douglas McGregor's *The Human Side of Enterprise*
Michael Porter's *Competitive Strategy: Creating and Sustaining Superior Performance*
John Kotter's *Leading Change*
C. K. Prahalad & Gary Hamel's *The Core Competence of the Corporation*

CRIMINOLOGY

Michelle Alexander's *The New Jim Crow: Mass Incarceration in the Age of Colorblindness*
Michael R. Gottfredson & Travis Hirschi's *A General Theory of Crime*
Richard Herrnstein & Charles A. Murray's *The Bell Curve: Intelligence and Class Structure in American Life*
Elizabeth Loftus's *Eyewitness Testimony*
Jay Macleod's *Ain't No Makin' It: Aspirations and Attainment in a Low-Income Neighborhood*
Philip Zimbardo's *The Lucifer Effect*

ECONOMICS

Janet Abu-Lughod's *Before European Hegemony*
Ha-Joon Chang's *Kicking Away the Ladder*
David Brion Davis's *The Problem of Slavery in the Age of Revolution*
Milton Friedman's *The Role of Monetary Policy*
Milton Friedman's *Capitalism and Freedom*
David Graeber's *Debt: the First 5000 Years*
Friedrich Hayek's *The Road to Serfdom*
Karen Ho's *Liquidated: An Ethnography of Wall Street*

The Macat Library By Discipline

John Maynard Keynes's *The General Theory of Employment, Interest and Money*
Charles P. Kindleberger's *Manias, Panics and Crashes*
Robert Lucas's *Why Doesn't Capital Flow from Rich to Poor Countries?*
Burton G. Malkiel's *A Random Walk Down Wall Street*
Thomas Robert Malthus's *An Essay on the Principle of Population*
Karl Marx's *Capital*
Thomas Piketty's *Capital in the Twenty-First Century*
Amartya Sen's *Development as Freedom*
Adam Smith's *The Wealth of Nations*
Nassim Nicholas Taleb's *The Black Swan: The Impact of the Highly Improbable*
Amos Tversky's & Daniel Kahneman's *Judgment under Uncertainty: Heuristics and Biases*
Mahbub Ul Haq's *Reflections on Human Development*
Max Weber's *The Protestant Ethic and the Spirit of Capitalism*

FEMINISM AND GENDER STUDIES

Judith Butler's *Gender Trouble*
Simone De Beauvoir's *The Second Sex*
Michel Foucault's *History of Sexuality*
Betty Friedan's *The Feminine Mystique*
Saba Mahmood's *The Politics of Piety: The Islamic Revival and the Feminist Subject*
Joan Wallach Scott's *Gender and the Politics of History*
Mary Wollstonecraft's *A Vindication of the Rights of Woman*
Virginia Woolf's *A Room of One's Own*

GEOGRAPHY

The Brundtland Report's *Our Common Future*
Rachel Carson's *Silent Spring*
Charles Darwin's *On the Origin of Species*
James Ferguson's *The Anti-Politics Machine*
Jane Jacobs's *The Death and Life of Great American Cities*
James Lovelock's *Gaia: A New Look at Life on Earth*
Amartya Sen's *Development as Freedom*
Mathis Wackernagel & William Rees's *Our Ecological Footprint*

HISTORY

Janet Abu-Lughod's *Before European Hegemony*
Benedict Anderson's *Imagined Communities*
Bernard Bailyn's *The Ideological Origins of the American Revolution*
Hanna Batatu's *The Old Social Classes And The Revolutionary Movements Of Iraq*
Christopher Browning's *Ordinary Men: Reserve Police Batallion 101 and the Final Solution in Poland*
Edmund Burke's *Reflections on the Revolution in France*
William Cronon's *Nature's Metropolis: Chicago And The Great West*
Alfred W. Crosby's *The Columbian Exchange*
Hamid Dabashi's *Iran: A People Interrupted*
David Brion Davis's *The Problem of Slavery in the Age of Revolution*
Nathalie Zemon Davis's *The Return of Martin Guerre*
Jared Diamond's *Guns, Germs & Steel: the Fate of Human Societies*
Frank Dikotter's *Mao's Great Famine*
John W Dower's *War Without Mercy: Race And Power In The Pacific War*
W. E. B. Du Bois's *The Souls of Black Folk*
Richard J. Evans's *In Defence of History*
Lucien Febvre's *The Problem of Unbelief in the 16th Century*
Sheila Fitzpatrick's *Everyday Stalinism*

Eric Foner's *Reconstruction: America's Unfinished Revolution, 1863-1877*
Michel Foucault's *Discipline and Punish*
Michel Foucault's *History of Sexuality*
Francis Fukuyama's *The End of History and the Last Man*
John Lewis Gaddis's *We Now Know: Rethinking Cold War History*
Ernest Gellner's *Nations and Nationalism*
Eugene Genovese's *Roll, Jordan, Roll: The World the Slaves Made*
Carlo Ginzburg's *The Night Battles*
Daniel Goldhagen's *Hitler's Willing Executioners*
Jack Goldstone's *Revolution and Rebellion in the Early Modern World*
Antonio Gramsci's *The Prison Notebooks*
Alexander Hamilton, John Jay & James Madison's *The Federalist Papers*
Christopher Hill's *The World Turned Upside Down*
Carole Hillenbrand's *The Crusades: Islamic Perspectives*
Thomas Hobbes's *Leviathan*
Eric Hobsbawm's *The Age Of Revolution*
John A. Hobson's *Imperialism: A Study*
Albert Hourani's *History of the Arab Peoples*
Samuel P. Huntington's *The Clash of Civilizations and the Remaking of World Order*
C. L. R. James's *The Black Jacobins*
Tony Judt's *Postwar: A History of Europe Since 1945*
Ernst Kantorowicz's *The King's Two Bodies: A Study in Medieval Political Theology*
Paul Kennedy's *The Rise and Fall of the Great Powers*
Ian Kershaw's *The "Hitler Myth": Image and Reality in the Third Reich*
John Maynard Keynes's *The General Theory of Employment, Interest and Money*
Charles P. Kindleberger's *Manias, Panics and Crashes*
Martin Luther King Jr's *Why We Can't Wait*
Henry Kissinger's *World Order: Reflections on the Character of Nations and the Course of History*
Thomas Kuhn's *The Structure of Scientific Revolutions*
Georges Lefebvre's *The Coming of the French Revolution*
John Locke's *Two Treatises of Government*
Niccolò Machiavelli's *The Prince*
Thomas Robert Malthus's *An Essay on the Principle of Population*
Mahmood Mamdani's *Citizen and Subject: Contemporary Africa And The Legacy Of Late Colonialism*
Karl Marx's *Capital*
Stanley Milgram's *Obedience to Authority*
John Stuart Mill's *On Liberty*
Thomas Paine's *Common Sense*
Thomas Paine's *Rights of Man*
Geoffrey Parker's *Global Crisis: War, Climate Change and Catastrophe in the Seventeenth Century*
Jonathan Riley-Smith's *The First Crusade and the Idea of Crusading*
Jean-Jacques Rousseau's *The Social Contract*
Joan Wallach Scott's *Gender and the Politics of History*
Theda Skocpol's *States and Social Revolutions*
Adam Smith's *The Wealth of Nations*
Timothy Snyder's *Bloodlands: Europe Between Hitler and Stalin*
Sun Tzu's *The Art of War*
Keith Thomas's *Religion and the Decline of Magic*
Thucydides's *The History of the Peloponnesian War*
Frederick Jackson Turner's *The Significance of the Frontier in American History*
Odd Arne Westad's *The Global Cold War: Third World Interventions And The Making Of Our Times*

LITERATURE

Chinua Achebe's *An Image of Africa: Racism in Conrad's Heart of Darkness*
Roland Barthes's *Mythologies*
Homi K. Bhabha's *The Location of Culture*
Judith Butler's *Gender Trouble*
Simone De Beauvoir's *The Second Sex*
Ferdinand De Saussure's *Course in General Linguistics*
T. S. Eliot's *The Sacred Wood: Essays on Poetry and Criticism*
Zora Neale Huston's *Characteristics of Negro Expression*
Toni Morrison's *Playing in the Dark: Whiteness in the American Literary Imagination*
Edward Said's *Orientalism*
Gayatri Chakravorty Spivak's *Can the Subaltern Speak?*
Mary Wollstonecraft's *A Vindication of the Rights of Women*
Virginia Woolf's *A Room of One's Own*

PHILOSOPHY

Elizabeth Anscombe's *Modern Moral Philosophy*
Hannah Arendt's *The Human Condition*
Aristotle's *Metaphysics*
Aristotle's *Nicomachean Ethics*
Edmund Gettier's *Is Justified True Belief Knowledge?*
Georg Wilhelm Friedrich Hegel's *Phenomenology of Spirit*
David Hume's *Dialogues Concerning Natural Religion*
David Hume's *The Enquiry for Human Understanding*
Immanuel Kant's *Religion within the Boundaries of Mere Reason*
Immanuel Kant's *Critique of Pure Reason*
Søren Kierkegaard's *The Sickness Unto Death*
Søren Kierkegaard's *Fear and Trembling*
C. S. Lewis's *The Abolition of Man*
Alasdair MacIntyre's *After Virtue*
Marcus Aurelius's *Meditations*
Friedrich Nietzsche's *On the Genealogy of Morality*
Friedrich Nietzsche's *Beyond Good and Evil*
Plato's *Republic*
Plato's *Symposium*
Jean-Jacques Rousseau's *The Social Contract*
Gilbert Ryle's *The Concept of Mind*
Baruch Spinoza's *Ethics*
Sun Tzu's *The Art of War*
Ludwig Wittgenstein's *Philosophical Investigations*

POLITICS

Benedict Anderson's *Imagined Communities*
Aristotle's *Politics*
Bernard Bailyn's *The Ideological Origins of the American Revolution*
Edmund Burke's *Reflections on the Revolution in France*
John C. Calhoun's *A Disquisition on Government*
Ha-Joon Chang's *Kicking Away the Ladder*
Hamid Dabashi's *Iran: A People Interrupted*
Hamid Dabashi's *Theology of Discontent: The Ideological Foundation of the Islamic Revolution in Iran*
Robert Dahl's *Democracy and its Critics*
Robert Dahl's *Who Governs?*
David Brion Davis's *The Problem of Slavery in the Age of Revolution*

Alexis De Tocqueville's *Democracy in America*
James Ferguson's *The Anti-Politics Machine*
Frank Dikotter's *Mao's Great Famine*
Sheila Fitzpatrick's *Everyday Stalinism*
Eric Foner's *Reconstruction: America's Unfinished Revolution, 1863-1877*
Milton Friedman's *Capitalism and Freedom*
Francis Fukuyama's *The End of History and the Last Man*
John Lewis Gaddis's *We Now Know: Rethinking Cold War History*
Ernest Gellner's *Nations and Nationalism*
David Graeber's *Debt: the First 5000 Years*
Antonio Gramsci's *The Prison Notebooks*
Alexander Hamilton, John Jay & James Madison's *The Federalist Papers*
Friedrich Hayek's *The Road to Serfdom*
Christopher Hill's *The World Turned Upside Down*
Thomas Hobbes's *Leviathan*
John A. Hobson's *Imperialism: A Study*
Samuel P. Huntington's *The Clash of Civilizations and the Remaking of World Order*
Tony Judt's *Postwar: A History of Europe Since 1945*
David C. Kang's *China Rising: Peace, Power and Order in East Asia*
Paul Kennedy's *The Rise and Fall of Great Powers*
Robert Keohane's *After Hegemony*
Martin Luther King Jr.'s *Why We Can't Wait*
Henry Kissinger's *World Order: Reflections on the Character of Nations and the Course of History*
John Locke's *Two Treatises of Government*
Niccolò Machiavelli's *The Prince*
Thomas Robert Malthus's *An Essay on the Principle of Population*
Mahmood Mamdani's *Citizen and Subject: Contemporary Africa And The Legacy Of Late Colonialism*
Karl Marx's *Capital*
John Stuart Mill's *On Liberty*
John Stuart Mill's *Utilitarianism*
Hans Morgenthau's *Politics Among Nations*
Thomas Paine's *Common Sense*
Thomas Paine's *Rights of Man*
Thomas Piketty's *Capital in the Twenty-First Century*
Robert D. Putman's *Bowling Alone*
John Rawls's *Theory of Justice*
Jean-Jacques Rousseau's *The Social Contract*
Theda Skocpol's *States and Social Revolutions*
Adam Smith's *The Wealth of Nations*
Sun Tzu's *The Art of War*
Henry David Thoreau's *Civil Disobedience*
Thucydides's *The History of the Peloponnesian War*
Kenneth Waltz's *Theory of International Politics*
Max Weber's *Politics as a Vocation*
Odd Arne Westad's *The Global Cold War: Third World Interventions And The Making Of Our Times*

POSTCOLONIAL STUDIES

Roland Barthes's *Mythologies*
Frantz Fanon's *Black Skin, White Masks*
Homi K. Bhabha's *The Location of Culture*
Gustavo Gutiérrez's *A Theology of Liberation*
Edward Said's *Orientalism*
Gayatri Chakravorty Spivak's *Can the Subaltern Speak?*

PSYCHOLOGY

Gordon Allport's *The Nature of Prejudice*
Alan Baddeley & Graham Hitch's *Aggression: A Social Learning Analysis*
Albert Bandura's *Aggression: A Social Learning Analysis*
Leon Festinger's *A Theory of Cognitive Dissonance*
Sigmund Freud's *The Interpretation of Dreams*
Betty Friedan's *The Feminine Mystique*
Michael R. Gottfredson & Travis Hirschi's *A General Theory of Crime*
Eric Hoffer's *The True Believer: Thoughts on the Nature of Mass Movements*
William James's *Principles of Psychology*
Elizabeth Loftus's *Eyewitness Testimony*
A. H. Maslow's *A Theory of Human Motivation*
Stanley Milgram's *Obedience to Authority*
Steven Pinker's *The Better Angels of Our Nature*
Oliver Sacks's *The Man Who Mistook His Wife For a Hat*
Richard Thaler & Cass Sunstein's *Nudge: Improving Decisions About Health, Wealth and Happiness*
Amos Tversky's *Judgment under Uncertainty: Heuristics and Biases*
Philip Zimbardo's *The Lucifer Effect*

SCIENCE

Rachel Carson's *Silent Spring*
William Cronon's *Nature's Metropolis: Chicago And The Great West*
Alfred W. Crosby's *The Columbian Exchange*
Charles Darwin's *On the Origin of Species*
Richard Dawkin's *The Selfish Gene*
Thomas Kuhn's *The Structure of Scientific Revolutions*
Geoffrey Parker's *Global Crisis: War, Climate Change and Catastrophe in the Seventeenth Century*
Mathis Wackernagel & William Rees's *Our Ecological Footprint*

SOCIOLOGY

Michelle Alexander's *The New Jim Crow: Mass Incarceration in the Age of Colorblindness*
Gordon Allport's *The Nature of Prejudice*
Albert Bandura's *Aggression: A Social Learning Analysis*
Hanna Batatu's *The Old Social Classes And The Revolutionary Movements Of Iraq*
Ha-Joon Chang's *Kicking Away the Ladder*
W. E. B. Du Bois's *The Souls of Black Folk*
Émile Durkheim's *On Suicide*
Frantz Fanon's *Black Skin, White Masks*
Frantz Fanon's *The Wretched of the Earth*
Eric Foner's *Reconstruction: America's Unfinished Revolution, 1863-1877*
Eugene Genovese's *Roll, Jordan, Roll: The World the Slaves Made*
Jack Goldstone's *Revolution and Rebellion in the Early Modern World*
Antonio Gramsci's *The Prison Notebooks*
Richard Herrnstein & Charles A Murray's *The Bell Curve: Intelligence and Class Structure in American Life*
Eric Hoffer's *The True Believer: Thoughts on the Nature of Mass Movements*
Jane Jacobs's *The Death and Life of Great American Cities*
Robert Lucas's *Why Doesn't Capital Flow from Rich to Poor Countries?*
Jay Macleod's *Ain't No Makin' It: Aspirations and Attainment in a Low Income Neighborhood*
Elaine May's *Homeward Bound: American Families in the Cold War Era*
Douglas McGregor's *The Human Side of Enterprise*
C. Wright Mills's *The Sociological Imagination*

Thomas Piketty's *Capital in the Twenty-First Century*
Robert D. Putman's *Bowling Alone*
David Riesman's *The Lonely Crowd: A Study of the Changing American Character*
Edward Said's *Orientalism*
Joan Wallach Scott's *Gender and the Politics of History*
Theda Skocpol's *States and Social Revolutions*
Max Weber's *The Protestant Ethic and the Spirit of Capitalism*

THEOLOGY

Augustine's *Confessions*
Benedict's *Rule of St Benedict*
Gustavo Gutiérrez's *A Theology of Liberation*
Carole Hillenbrand's *The Crusades: Islamic Perspectives*
David Hume's *Dialogues Concerning Natural Religion*
Immanuel Kant's *Religion within the Boundaries of Mere Reason*
Ernst Kantorowicz's *The King's Two Bodies: A Study in Medieval Political Theology*
Søren Kierkegaard's *The Sickness Unto Death*
C. S. Lewis's *The Abolition of Man*
Saba Mahmood's *The Politics of Piety: The Islamic Revival and the Feminist Subject*
Baruch Spinoza's *Ethics*
Keith Thomas's *Religion and the Decline of Magic*

COMING SOON

Chris Argyris's *The Individual and the Organisation*
Seyla Benhabib's *The Rights of Others*
Walter Benjamin's *The Work Of Art in the Age of Mechanical Reproduction*
John Berger's *Ways of Seeing*
Pierre Bourdieu's *Outline of a Theory of Practice*
Mary Douglas's *Purity and Danger*
Roland Dworkin's *Taking Rights Seriously*
James G. March's *Exploration and Exploitation in Organisational Learning*
Ikujiro Nonaka's *A Dynamic Theory of Organizational Knowledge Creation*
Griselda Pollock's *Vision and Difference*
Amartya Sen's *Inequality Re-Examined*
Susan Sontag's *On Photography*
Yasser Tabbaa's *The Transformation of Islamic Art*
Ludwig von Mises's *Theory of Money and Credit*

Macat Disciplines

Access the greatest ideas and thinkers across entire disciplines, including

AFRICANA STUDIES

Chinua Achebe's *An Image of Africa: Racism in Conrad's Heart of Darkness*

W. E. B. Du Bois's *The Souls of Black Folk*

Zora Neale Hurston's *Characteristics of Negro Expression*

Martin Luther King Jr.'s *Why We Can't Wait*

Toni Morrison's *Playing in the Dark: Whiteness in the American Literary Imagination*

Macat Disciplines

Access the greatest ideas and thinkers across entire disciplines, including

FEMINISM, GENDER AND QUEER STUDIES

Simone De Beauvoir's
The Second Sex

Michel Foucault's
History of Sexuality

Betty Friedan's
The Feminine Mystique

Saba Mahmood's
*The Politics of Piety:
The Islamic Revival and
the Feminist Subject*

Joan Wallach Scott's
*Gender and the
Politics of History*

Mary Wollstonecraft's
*A Vindication of the
Rights of Woman*

Virginia Woolf's
A Room of One's Own

Judith Butler's
Gender Trouble

Macat Disciplines

Access the greatest ideas and thinkers across entire disciplines, including

INEQUALITY

Ha-Joon Chang's, *Kicking Away the Ladder*

David Graeber's, *Debt: The First 5000 Years*

Robert E. Lucas's, *Why Doesn't Capital Flow from Rich To Poor Countries?*

Thomas Piketty's, *Capital in the Twenty-First Century*

Amartya Sen's, *Inequality Re-Examined*

Mahbub Ul Haq's, *Reflections on Human Development*

Macat analyses are available from all good bookshops and libraries.

Access hundreds of analyses through one, multimedia tool.

Join free for one month **library.macat.com**

Macat Disciplines

Access the greatest ideas and thinkers across entire disciplines, including

CRIMINOLOGY

Michelle Alexander's
*The New Jim Crow:
Mass Incarceration in the
Age of Colorblindness*

**Michael R. Gottfredson
& Travis Hirschi's**
A General Theory of Crime

Elizabeth Loftus's
Eyewitness Testimony

**Richard Herrnstein
& Charles A. Murray's**
*The Bell Curve: Intelligence and
Class Structure in American Life*

Jay Macleod's
*Ain't No Makin' It:
Aspirations and Attainment in a
Low-Income Neighborhood*

Philip Zimbardo's
The Lucifer Effect

Macat analyses are available from all good bookshops and libraries.

Access hundreds of analyses through one, multimedia tool.
Join free for one month **library.macat.com**

Macat Disciplines

Access the greatest ideas and thinkers across entire disciplines, including

Postcolonial Studies

Roland Barthes's *Mythologies*
Frantz Fanon's *Black Skin, White Masks*
Homi K. Bhabha's *The Location of Culture*
Gustavo Gutiérrez's *A Theology of Liberation*
Edward Said's *Orientalism*
Gayatri Chakravorty Spivak's *Can the Subaltern Speak?*

Macat analyses are available from all good bookshops and libraries.

Access hundreds of analyses through one, multimedia tool.
Join free for one month **library.macat.com**

Macat Pairs

Analyse historical and modern issues from opposite sides of an argument. Pairs include:

HOW TO RUN AN ECONOMY

John Maynard Keynes's
The General Theory OF Employment, Interest and Money

Classical economics suggests that market economies are self-correcting in times of recession or depression, and tend toward full employment and output. But English economist John Maynard Keynes disagrees.

In his ground-breaking 1936 study *The General Theory*, Keynes argues that traditional economics has misunderstood the causes of unemployment. Employment is not determined by the price of labor; it is directly linked to demand. Keynes believes market economies are by nature unstable, and so require government intervention. Spurred on by the social catastrophe of the Great Depression of the 1930s, he sets out to revolutionize the way the world thinks

Milton Friedman's
The Role of Monetary Policy

Friedman's 1968 paper changed the course of economic theory. In just 17 pages, he demolished existing theory and outlined an effective alternate monetary policy designed to secure 'high employment, stable prices and rapid growth.'

Friedman demonstrated that monetary policy plays a vital role in broader economic stability and argued that economists got their monetary policy wrong in the 1950s and 1960s by misunderstanding the relationship between inflation and unemployment. Previous generations of economists had believed that governments could permanently decrease unemployment by permitting inflation—and vice versa. Friedman's most original contribution was to show that this supposed trade-off is an illusion that only works in the short term.

Macat analyses are available from all good bookshops and libraries.

Access hundreds of analyses through one, multimedia tool.
Join free for one month **library.macat.com**

Macat Disciplines

Access the greatest ideas and thinkers across entire disciplines, including

THE FUTURE OF DEMOCRACY

Robert A. Dahl's, *Democracy and Its Critics*
Robert A. Dahl's, *Who Governs?*
Alexis De Toqueville's, *Democracy in America*
Niccolò Machiavelli's, *The Prince*
John Stuart Mill's, *On Liberty*
Robert D. Putnam's, *Bowling Alone*
Jean-Jacques Rousseau's, *The Social Contract*
Henry David Thoreau's, *Civil Disobedience*

Macat Pairs

Analyse historical and modern issues from opposite sides of an argument. Pairs include:

RACE AND IDENTITY

Zora Neale Hurston's
Characteristics of Negro Expression

Using material collected on anthropological expeditions to the South, Zora Neale Hurston explains how expression in African American culture in the early twentieth century departs from the art of white America. At the time, African American art was often criticized for copying white culture. For Hurston, this criticism misunderstood how art works. European tradition views art as something fixed. But Hurston describes a creative process that is alive, ever-changing, and largely improvisational. She maintains that African American art works through a process called 'mimicry'—where an imitated object or verbal pattern, for example, is reshaped and altered until it becomes something new, novel—and worthy of attention.

Frantz Fanon's
Black Skin, White Masks

Black Skin, White Masks offers a radical analysis of the psychological effects of colonization on the colonized.

Fanon witnessed the effects of colonization first hand both in his birthplace, Martinique, and again later in life when he worked as a psychiatrist in another French colony, Algeria. His text is uncompromising in form and argument. He dissects the dehumanizing effects of colonialism, arguing that it destroys the native sense of identity, forcing people to adapt to an alien set of values—including a core belief that they are inferior. This results in deep psychological trauma.

Fanon's work played a pivotal role in the civil rights movements of the 1960s.

Macat Pairs

Analyse historical and modern issues from opposite sides of an argument. Pairs include:

INTERNATIONAL RELATIONS IN THE 21ST CENTURY

Samuel P. Huntington's
The Clash of Civilisations

In his highly influential 1996 book, Huntington offers a vision of a post-Cold War world in which conflict takes place not between competing ideologies but between cultures. The worst clash, he argues, will be between the Islamic world and the West: the West's arrogance and belief that its culture is a "gift" to the world will come into conflict with Islam's obstinacy and concern that its culture is under attack from a morally decadent "other."

Clash inspired much debate between different political schools of thought. But its greatest impact came in helping define American foreign policy in the wake of the 2001 terrorist attacks in New York and Washington.

Francis Fukuyama's
The End of History and the Last Man

Published in 1992, *The End of History and the Last Man* argues that capitalist democracy is the final destination for all societies. Fukuyama believed democracy triumphed during the Cold War because it lacks the "fundamental contradictions" inherent in communism and satisfies our yearning for freedom and equality. Democracy therefore marks the endpoint in the evolution of ideology, and so the "end of history." There will still be "events," but no fundamental change in ideology.

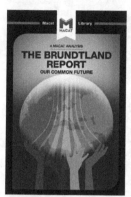

Printed in the United States
by Baker & Taylor Publisher Services